Production of Clean Hydrogen by Electrochemical Reforming of Oxygenated Organic Compounds

Hydrogen and Fuel Cells Primers

Production of Clean Hydrogen by Electrochemical Reforming of Oxygenated Organic Compounds

CLAUDE LAMY

European Institute of Membranes, IEM, UMR n°5635, CNRS-University of Montpellier, Montpellier, France

CHRISTOPHE COUTANCEAU

Institute of Material and Environment Chemistry of Poitiers, IC2MP, UMR n°7285, CNRS-University of Poitiers, Poitiers, France

STÉVE BARANTON

Institute of Material and Environment Chemistry of Poitiers, IC2MP, UMR n°7285, CNRS-University of Poitiers, Poitiers, France

Series Editor

BRUNO G. POLLET

Professor of Renewable Energy, Norwegian University of Science and Technology (NTNU), Norway

ELSEVIER

ACADEMIC PRESS

An imprint of Elsevier

Academic Press is an imprint of Elsevier
125 London Wall, London EC2Y 5AS, United Kingdom
525 B Street, Suite 1650, San Diego, CA 92101, United States
50 Hampshire Street, 5th Floor, Cambridge, MA 02139, United States
The Boulevard, Langford Lane, Kidlington, Oxford OX5 1GB, United Kingdom

Notices

Knowledge and best practice in this field are constantly changing. As new research and experience broaden our understanding, changes in research methods, professional practices, or medical treatment may become necessary.

Practitioners and researchers must always rely on their own experience and knowledge in evaluating and using any information, methods, compounds, or experiments described herein. In using such information or methods they should be mindful of their own safety and the safety of others, including parties for whom they have a professional responsibility.

To the fullest extent of the law, neither the Publisher nor the authors, contributors, or editors, assume any liability for any injury and/or damage to persons or property as a matter of products liability, negligence or otherwise, or from any use or operation of any methods, products, instructions, or ideas contained in the material herein.

British Library Cataloguing-in-Publication Data

A catalogue record for this book is available from the British Library

Library of Congress Cataloging-in-Publication Data

A catalog record for this book is available from the Library of Congress

ISBN: 978-0-12-821500-5

Publisher: Brian Romer
Acquisition Editor: Graham Nisbet
Editorial Project Manager: Ali Afzal-Khan
Production Project Manager: Surya Narayanan Jayachandran
Cover Designer: Christian J. Bilbow

Typeset by MPS Limited, Chennai, India

Contents

Acknowledgments

The authors greatly acknowledge the CNRS Research Grouping HySPàC (GDR CNRS n°3652), and CL acknowledges the European Membrane Institute of Montpellier (UMR CNRS n°5635) and the Chemistry Institute of the "Centre National de la Recherche Scientifique" (CNRS) for supporting this work; CC and SB thank the "PEPS energie" from CNRS, the European communities (FEDER, through the Econat project), the "Région Nouvelle Aquitaine," and the "Agence Nationale de la Recherche" (Project ECO-Plan, grant number ANR 16-CE29-0007) for financial supports.

List of abbreviations

AA	Acetic acid
AAL	Acetaldehyde
AAS	Atomic absorption spectroscopy
AEC	Alkaline electrolysis cell
ATR	Autothermal reforming
CPL	Carbon porous layer
DAFC	Direct alcohol fuel cell
DEFC	Direct ethanol fuel cell
DEMS	Differential electrochemical mass spectrometry
DFAFC	Direct formic acid fuel cell
DFT	Density Functional Theory
DMFC	Direct methanol fuel cell
DOE	Department of Energy
EDX	Energy dispersive X-ray spectrometry
FTIRS	Fourier Transform Infrared Reflectance Spectroscopy
GDE	Gas Diffusion Electrode
HER	Hydrogen evolution reaction
HOR	Hydrogen oxidation reaction
HPLC	High-performance liquid chromatography
ICP-OES	Inductively coupled plasma-optical emission spectroscopy
MEA	Membrane electrode assembly
MSE	Mercury/mercurous sulfate electrode
OER	Oxygen evolution reaction
ORR	Oxygen reduction reaction
PEM	Proton exchange membrane
PEMEC	Proton exchange membrane electrolysis cell
PEMFC	Proton exchange membrane fuel cell
PrOx	Preferential oxidation
rds	Rate determining step
RHE	Reversible hydrogen electrode
rpm	Round per minute
SHE	Standard hydrogen electrode
SMR	Steam methane reforming
SPAIRS	Single potential alteration infrared spectroscopy
SR	Steam reforming
SSE	Silver/silver chloride electrode
TEM	Transmission electron microscopy
TGA	Thermo-gravimetric analysis
XPS	X-ray photoelectron spectroscopy
XRD	X-ray diffraction

CHAPTER ONE

Introduction

Contents

In the near future, hydrogen is considered as a key compound for energy production and storage, besides its use in chemical industries. Its implication as an energy career is expected to deal with the environmental issues to respond to the growing energy demand and, at the same time, to allow economic expansion. Hydrogen is widely used for the production of fertilizers and chemicals; more recently, one of the main applications for hydrogen energy concerns feeding fuel cells [1]. These systems display very high energy efficiencies compared with internal combustion engines [2], and more and more car manufacturers are already proposing or will be proposing fuel-cell electric vehicles on the market in the next years. Therefore, the hydrogen demand will inevitably increase, and the development of methods for delivering hydrogen safely is becoming of paramount importance. Currently, 96% of the hydrogen is produced from fossil sources, mainly from natural gas by steam methane reforming. Indeed, the delivery of hydrogen mainly leans on processes such as reforming of natural gas (c.48%), oil (c.30%), and gasification or reforming of coal (c.18%) [3]. Such processes based on fossil resources obviously make hydrogen production not yet "green" or at least not yet sustainable.

One alternative for producing pure hydrogen is the electrolysis of water. The splitting of the water molecule into low-temperature electrolysis cells (such as a proton exchange membrane electrolysis cell [PEMEC] or an alkaline electrolysis cell) leading to oxygen at the anode and hydrogen at the cathode is indeed currently extensively studied [4,5]. Historically alkaline technology was the most studied and developed process because the environment is less corrosive than an acidic one [6]. However, the development of highly active Ir- or Ru-based dimensionally stable anodes allowed the proton exchange membrane electrolysis technology reaching higher efficiency than the alkaline technology [4,7]. Furthermore, in this case, renewable intermittent primary energy sources,

Production of Clean Hydrogen by Electrochemical Reforming of Oxygenated Organic Compounds.
DOI: https://doi.org/10.1016/B978-0-12-821500-5.00001-1

such as wind, solar, tidal, and so on, can be used to make this technology greener. However, one of the major problems of these technologies comes from the high electrical energy needed to dissociate the water molecule ($\Delta H^0 = 285.8$ kJ mol^{-1} and $\Delta G^0 = 236.8$ kJ mol^{-1}), which leads to a very high standard cell voltage of 1.23 V. Moreover, the oxygen evolution reaction occurs with a very low kinetics, so that high overvoltages are necessary to reach acceptable current densities of at least 1 A cm^{-2} (i.e., to obtain a high hydrogen production rate at the cathode). Thus, cell voltages of 1.8–2.0 V are generally required for the significant hydrogen production. These overvoltages are responsible for the high cost of hydrogen production via the water electrolysis process, not only in terms of electrical energy required but also in terms of materials for the anode, for example, Ru- and Ir-based catalysts and corrosion-resistant support based on titanium.

Liquid alcohols, polyols, and sugars from the biomass could serve as a bridge between hydrocarbon sources and gaseous H_2. They own high energy densities and can be easily stored and distributed through the existing gasoline infrastructure. It has also been claimed that alcohols are the next liquid fuels to use after the depletion of petroleum resources [8]. Alcohols can be used to produce electricity in direct fuel cells or mechanical energy in internal combustion engines, such as in the ProAlcohol Plan developed in Brazil. However, the performances, between a few tens and a few hundreds of mW cm^{-2}, of a direct alcohol fuel cell (DAFC) remain relatively low [9,10], compared with those obtained by a hydrogen proton exchange membrane fuel cell (PEMFC) reaching more than 1 W cm^{-2}. The DAFC performances are low due to the low kinetics of the electro-oxidation reaction of organic molecules and to the poisoning of the catalyst surface by adsorbed intermediate species at low potentials [11,12]. Moreover, the difficulty to activate alcohol oxidation at low temperatures, below 100°C, necessitates high loadings of catalysts, for example, several mg cm^{-2}, at the anode of the fuel cell in order to achieve acceptable performances.

Furthermore, very clean hydrogen is required to feed a PEMFC. Indeed, the platinum-based catalysts used at the anode of the fuel cell can be easily poisoned through the strong adsorption of poisons, such as CO, sulfur, and others, which leads to a rapid and drastic decrease in performances [13,14]. Clean H_2 can be produced from alcohol by its oxidation in an electrochemical reformer or electrolyzer and can be used in different devices such as fuel cells, resulting in an improvement in the overall system performance. Electrochemical reforming or electrolysis is a process in which an organic compound such as alcohol is electro-oxidized to form H_2. Electrical power is required to split the chemically bonded species. In a PEMEC, the electrochemical reformer consists of two electrodes separated by an ion-exchange membrane, which form the so-called membrane electrode assembly (MEA). The role of the membrane will be to ensure the ionic conductivity between both anodic and cathodic compartments and to separate the reaction products (H_2 at the cathode and CO_2 and organic oxidation products at the anode), leading to the production of clean hydrogen. According to thermodynamic data (see Section 2.1), the oxidation of many alcohols needs smaller electrical energy (ΔG^+) than that of water splitting and leads to very low standard potentials, which is close to that of hydrogen. Their oxidation at the catalytic anode of an electrolysis cell is expected to occur at lower overvoltages, thus decreasing the working cell voltage, compared with that of a water electrolysis cell, and further to save energy for the production of clean hydrogen.

Methanol is the simplest alcohol and displays high reactivity. It is a good candidate that has been studied for the hydrogen production by its electrolysis for more than 17 years [15–18]. Formic acid is an interesting compound because it is very reactive and its oxidation leads to one molecule of H_2 and one molecule of CO_2, that is, $HCOOH \rightarrow H_2 + CO_2$. Dimethoxymethane ($CH_3-O-CH_2-O-CH_3$) is also a good hydrogen carrier and is expected to be reactive because in acidic medium it is hydrolyzed back to formaldehyde and methanol. However, methanol is a toxic compound, and all these chemicals are mainly produced from fossil resources, for example, by partial oxidation of methane. Ethanol can be a biosourced compound (bio-ethanol) from sugar fermentation. It has been shown to be less reactive than methanol in acidic electrolysis cell due to the presence of the C–C bond, which is difficult to cleave at low temperatures, that is, below 100°C. This lack of reactivity in acidic media limits the electrolysis cell efficiency. Alcohols are known to be more reactive in alkaline media than in acidic media [19,20]. Low platinum-loaded electrodes or even platinum-free electrodes can be used for the electrooxidation of alcohol in alkaline media [21–24]. However, the efficiency of catalysts for the C–C bond breaking is lower in such media. This property can be interesting if using wasted by-products from industrial processes. For example, the trans-esterification (methanolysis) reaction of vegetable oil to produce biodiesel oil leads to c.10 wt.% of glycerol as a side product [25,26]. In the case of bio-ethanol production by anaerobic fermentation, the significant amount of glycerol is also formed, that is, several wt.% [27]. The oxidation of glycerol in an electrolysis cell can lead to the coproduction of hydrogen at the cathode and C3 value-added compounds at the anode [28]. Other biosourced compounds, such as cellulose and hemicellulose, represent very important carbon resources, of which few are currently valorized. The depolymerization of these polysaccharides by nonthermal activation processes leads to the formation of water-soluble polysaccharides and monosaccharides [29–31], which can be further oxidized in an electrolysis cell to produce hydrogen at the cathode and reaction products at the anode. The oxidation of glucose and xylose in acidic medium is deeper than that in alkaline one, leading to C–C bond breaking and degradation products. In alkaline medium, it is possible to control the oxidation reaction to obtain value-added compounds with very high selectivity, such as C6 and C5 carboxylic acids, for example, gluconate and xylonate, respectively. It has indeed been recently demonstrated that the use of biomass (bio-derived alcohols and polyols, sugars, etc.) allowed increasing the efficiency of the electrolysis process for the H_2 production, decreasing the energy consumption, and producing value-added compounds [32].

In this book chapter, we will first present and explain the principle of hydrogen production by the electrocatalytic oxidation of organic compounds in a PEMEC. Then a comprehensive review of the recent and most important achievements concerning the hydrogen production by electroreforming of organic molecules will be presented. The hydrogen production by the electrocatalytic oxidation of low-weight compounds, such as HCOOH, MeOH, and EtOH, and biomass-derived compounds, such as glycerol and saccharides, will be particularly discussed in terms of thermodynamics, kinetics, and electrolysis cell performances.

References

[1] United States Department of Energy, Report of the hydrogen production expert panel to HTAC. <http://www.hydrogen.energy.gov/pdfs/hpep_report_2013.pdf>, 2013 (accessed 26.04.18).

[2] S.E. Wright, Comparison of the theoretical performance potential of fuel cells and heat engines, Renew. Energy 29 (2004) 179–195. Available from: https://doi.org/10.1016/S0960-1481(03)00191-5.

[3] C.M. Kalamaras, A.M. Efstathiou, Hydrogen production technologies: current state and future developments, Conf. Pap. Energy (2013) 1–9. Available from: https://doi.org/10.1155/2013/690627.

[4] M. Carmo, D.L. Fritz, J. Mergel, D. Stolten, A comprehensive review on PEM Water electrolysis, Int. J. Hydrogen Energy 38 (2013) 4901–4934. Available from: https://doi.org/10.1016/j.ijhydene.2013.01.151.

[5] M. Gong, D.Y. Wang, C.C. Chen, B.J. Hwang, H. Dai, A mini review on nickel-based electrocatalysts for alkaline hydrogen evolution reaction, Nano Res. 16 (2016) 28–46. Available from: https://doi.org/10.1007/s12274-015-0965-x.

[6] K. Zeng, D.K. Zhang, Recent progress in alkaline water electrolysis for hydrogen production and applications, Prog. Energy Combust. Sci. 36 (2010) 307–326. Available from: https://doi.org/10.1016/j.pecs.2009.11.002.

[7] A. Goñi-Urtiaga, D. Presvytes, K. Scott, Solid acids as electrolyte materials for proton exchange membrane (PEM) electrolysis: review, Int. J. Hydrogen Energy 37 (2012) 3358–3372. Available from: https://doi.org/10.1016/j.ijhydene.2011.09.152.

[8] C. Lamy, J.M. Léger, S. Srinivasan, Direct methanol fuel cells: from a twentieth century electrochemist's dream to a twenty-first century emerging technology, in: J.M. Bockris, B.E. Conway, R.E. White (Eds.), Modern Aspects of Electrochemistry, vol. 34, Kluwer Academic/Plenum Publishers, New York, 2001, pp. 53–118. Available from: https://doi.org/10.1007/0-306-46923-5_3.

[9] P. Joghee, J.N. Malik, S. Pylypenko, R. O'Hayre, Review on direct methanol fuel cells – in the perspective of energy and sustainability, MRS Energy Sustain.: Rev. J. 23 (2015) 1–31. Available from: https://doi.org/10.1557/mre.2015.4.

[10] M.Z.F. Kamarudin, S.K. Kamarudin, M.S. Masdar, W.R.W. Dau, Review: direct ethanol fuel cells, Int. J. Hydrogen Energy 38 (2013) 9438–9453. Available from: https://doi.org/10.1016/j.ijhydene.2012.07.059.

[11] C. Lamy, A. Lima, V. LeRhun, F. Delime, C. Coutanceau, J.M. Léger, Recent advances in the development of direct alcohol fuel cells (DAFC), J. Power Sources 105 (2002) 283–296. Available from: https://doi.org/10.1016/S0378-7753(01)00954-5.

[12] C. Lamy, S. Rousseau, E.M. Belgsir, C. Coutanceau, J.M. Léger, Recent progress in the direct ethanol fuel cell: development of new platinum–tin electrocatalysts, Electrochim. Acta 49 (2004) 3901–3908. Available from: https://doi.org/10.1016/j.electacta.2004.01.078.

[13] J.J. Baschuk, X. Li, Carbon monoxide poisoning of proton exchange membrane fuel cells, Int. J. Energy Res. 25 (2001) 695–713. Available from: https://doi.org/10.1002/er.713.

[14] V.A. Sethuraman, J.W. Weidner, Analysis of sulfur poisoning on a PEM fuel cell electrode, Electrochim. Acta 55 (2010) 5683–5694. Available from: https://doi.org/10.1016/j.electacta.2010.05.004.

[15] B. Jeffries-Nakamura, S.R. Narayanan, T.I. Valdez, W. Chun, Making hydrogen by electrolysis of methanol, NASA Tech. Brief 6 (2002) 26.

[16] S.R. Narayanan, W. Chun, B. Jeffries-Nakamura, T.I. Valdez. Hydrogen generation by electrolysis of aqueous organic solutions. US Patent 6299744 (2001); US Patent 6368492 (2002); US Patent 6643284 (2001); US Patent 6533919 (2003); US Patent 0226763 (2003); US Patent 7056428 B2 (2006).

[17] T. Take, K. Tsurutani, M. Umeda, Hydrogen production by methanol–water solution electrolysis, J. Power Sources 164 (2007) 9–16. Available from: https://doi.org/10.1016/j.jpowsour.2006.10.011.

[18] S. Uhm, H. Jeon, T.J. Kim, J. Lee, Clean hydrogen production from methanol–water solutions via power saved electrolytic reforming process, J. Power Sources 198 (2012) 218–222. Available from: https://doi.org/10.1016/j.jpowsour.2011.09.083.

[19] J. Prabhuram, R. Manoharan, Investigation of methanol oxidation on unsupported platinum electrodes in strong alkali and strong acid, J. Power Sources 74 (1998) 54–61. Available from: https://doi.org/10.1016/S0378-7753(98)00012-3.

[20] A.V. Tripkovic, K.D. Popovic, B.N. Grgur, B. Blizanac, P.N. Ross, N.M. Markovic, Methanol electrooxidation on supported Pt and PtRu catalysts in acid and alkaline solutions, Electrochim. Acta 47 (2002) 3707–3714. Available from: https://doi.org/10.1016/S0013-4686(02)00340-7.
[21] C. Bianchini, P.K. Shen, Palladium-based electrocatalysts for alcohol oxidation in half cells and in direct alcohol fuel cells, Chem. Rev. 109 (2009) 4183–4206. Available from: https://doi.org/10.1021/cr9000995.
[22] M. Simões, S. Baranton, C. Coutanceau, Electrochemical valorisation of glycerol, Chem. Sus. Chem. 5 (2012) 2106–2124. Available from: https://doi.org/10.1002/cssc.201200335.
[23] Y. Holade, C. Morais, K. Servat, T.W. Napporn, K.B. Kokoh, Toward the electrochemical valorization of glycerol: Fourier transform infrared spectroscopic and chromatographic studies, ACS Catal. 3 (2013) 2403–2411. Available from: https://doi.org/10.1021/cs400559d.
[24] C. Xu, Y. Hu, J. Rong, S.P. Jiang, Y. Liu, Ni hollow spheres as catalysts for methanol and ethanol electrooxidation, Electrochem. Comm. 9 (2007) 2009–2012. Available from: https://doi.org/10.1016/j.elecom.2007.05.028.
[25] J.M. Clacens, Y. Pouilloux, J. Barrault, Selective etherification of glycerol to polyglycerols over impregnated basic MCM-41 type mesoporous catalysts, Appl. Catal. A: Gen. 227 (2002) 181–190. Available from: https://doi.org/10.1016/S0926-860X(01)00920-6.
[26] M.A. Dasari, P.P. Kiatsimkul, W.R. Sutterlin, G.J. Suppes, Low-pressure hydrogenolysis of glycerol to propylene glycol, Appl Catal A: Gen. 281 (2005) 225–231. Available from: https://doi.org/10.1016/j.apcata.2004.11.033.
[27] A.S. Aldiguier, S. Alfenore, X. Cameleyre, G. Goma, J.L. Uribelarrea, S.E. Guillouet, et al., Synergistic temperature and ethanol effect on Saccharomyces cerevisiae dynamic behaviour in ethanol bio-fuel production, Bioprocess. Biosyst. Eng. 26 (2004) 217–222. Available from: https://doi.org/10.1007/s00449-004-0352-6.
[28] J. Gonzalez-Cobos, S. Baranton, C. Coutanceau, Development of bismuth-modified PtPd nanocatalysts for the electrochemical reforming of polyols into hydrogen and value-added chemicals, Chem. Electro. Chem. 3 (2016) 1694–1704. Available from: https://doi.org/10.1002/celc.201600147.
[29] M. Benoit, A. Rodrigues, Q. Zhang, E. Fourré, K. De Oliveira Vigier, J.M. Tatibouët, et al., Depolymerization of cellulose assisted by a nonthermal atmospheric plasma, Angew. Chem. Int. Ed. 50 (2011) 8964–8967. Available from: https://doi.org/10.1002/anie.201104123.
[30] M. Benoit, A. Rodrigues, K. De Oliveira Vigier, E. Fourré, J. Barrault, J.M. Tatibouët, et al., Combination of ball-milling and non-thermal atmospheric plasma as physical treatments for the saccharification of microcrystalline cellulose, Green Chem. 14 (2012) 2212–2215. Available from: https://doi.org/10.1039/C2GC35710K.
[31] Q. Zhang, M. Benoit, K. De Oliveira Vigier, J. Barrault, F. Jérôme, Pretreatment of microcrystalline cellulose by ultrasounds: effect of particle size in the heterogeneously-catalyzed hydrolysis of cellulose to glucose, Green Chem. 15 (2013) 963–969. Available from: https://doi.org/10.1039/C3GC36643J.
[32] Y.X. Chen, A. Lavacchi, H.A. Miller, M. Bevilacqua, J. Filippi, M. Innocenti, et al., Nanotechnology makes biomass electrolysis more energy efficient than water electrolysis, Nat. Commun. 5 (2014) 4036. Available from: https://doi.org/10.1038/ncomms5036.

CHAPTER TWO

Principle of hydrogen production by electrocatalytic oxidation of organic compounds in a proton exchange membrane electrolysis cell

Contents

The electrochemical decomposition of an organic compound, $C_xH_yO_z$, for producing clean hydrogen can be performed in a proton exchange membrane electrolysis cell (PEMEC) of configuration similar to that of a water electrolysis cell (Fig. 2.1).

In the case of water electrolysis, the anodic compartment of the electrochemical cell is fed with water, under liquid or gaseous state, which is oxidized leading to the production of oxygen and protons, according to reaction (2.1):

$$H_2O \rightarrow \tfrac{1}{2}\, O_2 + 2\, H^+ + 2\, e^- \quad \text{anodic reaction} \tag{2.1}$$

Oxygen evolves in the gaseous phase, whereas the electrons circulate in the external circuit and protons cross-over the membrane, reaching the cathodic compartment where they are reduced by the electrons coming from the external circuit, thus producing hydrogen, as follows:

$$2\, H^+ + 2\, e^- \rightarrow H_2 \quad \text{cathodic reaction} \tag{2.2}$$

The electrical balance between reactions (2.1) and (2.2) corresponds to the electrochemical splitting of water into hydrogen and oxygen according to the overall reaction:

Production of Clean Hydrogen by Electrochemical Reforming of Oxygenated Organic Compounds.
DOI: https://doi.org/10.1016/B978-0-12-821500-5.00002-3

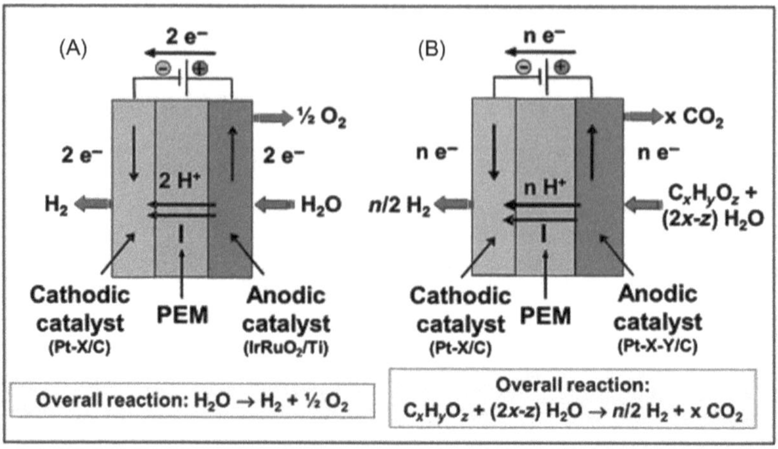

Figure 2.1 Principle of the electrochemical decomposition, in a proton exchange membrane electrolysis cell, of (A) water and (B) an organic compound $C_xH_yO_z$.

$$H_2O \rightarrow H_2 + \tfrac{1}{2}\,O_2 \quad \text{overall reaction} \tag{2.3}$$

with the following data under standard conditions ($T = 25°C$, $p = 1$ bar, liquid water):

$$\Delta H^0 = +\,285.8\ \text{kJ(mole H}_2\text{O)}^{-1} \ \text{and} \ \Delta G^0 = +\,237.2\ \text{kJ(mole H}_2\text{O)}^{-1}$$

These electrochemical processes are exactly the reverse of those occurring in a fuel cell with an acid electrolyte, for example, phosphoric acid, or a proton exchange membrane (PEM). For a PEM device, such as an electrolyzer or a fuel cell, the electrocatalysts at the negative electrode for the hydrogen evolution reaction (HER) or hydrogen oxidation reaction (HOR) are similar, that is, Pt/C nanoparticles, since the HER/HOR is a quasi-reversible reaction. Conversely, the electrocatalysts at the positive electrode are different, that is, Pt-based nanoparticles dispersed on the large surface area carbon powder, for example, Vulcan XC72R, for the oxygen reduction reaction (ORR) and Ir-based oxides dispersed on a corrosion-proof support, for example, a Ti mesh or foam, for the oxygen evolution reaction (OER).

In the case of an organic compound, $C_xH_yO_z$, its complete electro-oxidation in the anodic compartment of the electrolysis cell is realized with a suitable electrocatalyst, which is specific for the considered compound, leading to carbon dioxide and protons, according to reaction (2.4):

$$C_xH_yO_z + (2x - z)H_2O \rightarrow x\,CO_2 + n(H^+ + e^-) \quad \text{anodic reaction} \tag{2.4}$$

where $n = 4x + y - 2z$ is the number of exchanged electrons.

Then the protons produced cross-over the PEM and are reduced to molecular hydrogen by a convenient catalyst in the cathodic compartment by the electrons coming from the external circuit according to reaction (2.5):

$$n(H^+ + e^-) \rightarrow (n/2)H_2 \quad \text{cathodic reaction} \tag{2.5}$$

The electrical balance between reactions (2.4) and (2.5) leads to the following overall reaction:

$$C_xH_yO_z + (2\,x - z)H_2O \rightarrow x\,CO_2 + (n/2)H_2 \quad \text{overall reaction} \tag{2.6}$$

Reaction (2.6) corresponds to the electrochemical reforming of $C_xH_yO_z$ producing hydrogen and carbon dioxide by its complete oxidation.

The electrochemical decomposition of $C_xH_yO_z$, according to the overall reaction (2.6), is thus similar to a steam reforming process, but this reaction can occur at relatively low temperatures (25°C − 85°C) instead of the high temperatures (600°C − 800°C) encountered in steam reforming [1−5].

2.1 Thermodynamic considerations

Knowing the thermodynamic data of the formation of H_2O, CO_2, and $C_xH_yO_z$ under standard conditions, usually the liquid state at 25°C, that is, ΔH_f^0 and ΔG_f^0, one may calculate the thermodynamic data ΔH^+ and ΔG^+ associated with reaction (2.4) or (2.6) under standard conditions:

$$\Delta H^+ = x\Delta H^0_{CO_2} - (2x - z)\Delta H^0_{H_2O} - \Delta H_f^0$$

$$\Delta G^+ = x\Delta G^0_{CO_2} - (2x - z)\Delta G^0_{H_2O} - \Delta G_f^0$$

Since most of the electroreactive organic compounds can be used as fuels in a direct oxidation fuel cell (DOFC), the thermodynamic data of their complete oxidation corresponding to the overall reaction in the fuel cell device, that is,

$$C_xH_yO_z + (x + y/4 - z/2)O_2 \rightarrow x\,CO_2 + y/2\,H_2O \quad \text{overall reaction} \tag{2.7}$$

can be estimated from the combustion data of the compounds, ΔH^0_{FC} and ΔG^0_{FC}, in the fuel cell. Thus, the thermodynamic data of the electrochemical reforming reaction can be related to those of the combustion reaction of H_2 and $C_xH_yO_z$ as follows:

$$\Delta H^+ = \Delta H^0_{C_xH_yO_z,FC} - (n/2)\Delta H^0_{H_2,FC}$$

$$\Delta G^+ = \Delta G^0_{C_xH_yO_z,FC} - (n/2)\Delta G^0_{H_2,FC}$$

For low-weight organic compounds, the knowledge of their thermodynamic data of formation allowed us to calculate $\Delta h^+ = \Delta H^+/(2x + y/2 - z)$ corresponding to the total external energy needed to produce one mole of hydrogen and $E_a^+ = \Delta G^+/((4x + y - 2z)F)$, the minimum anode potential necessary to decompose the organic compound through its electrocatalytic oxidation. Since the cathode potential E_c^- associated with the HER (2.2) or (2.5) is close to 0 V versus the standard hydrogen electrode (SHE), using the Pt/C electrode in contact with the hydrogen contained in the cathodic compartment as a reference electrode, the minimum cell voltage to be applied to the electrolysis cell to decompose the organic compound will be given as follows:

$$U^0_{cell} = E_a^+ - E_c^- \approx E_a^+$$

For many organic compounds, this theoretical cell voltage (Table 2.1) is very small compared with that of the electrolytic decomposition of water, that is, $U^0_{cell} \approx 1.23$ V under standard conditions, so that the theoretical electrical energy, proportional to U^0_{cell}, to be provided by the external power source, to produce one hydrogen mole by the electro-oxidation of $C_xH_yO_z$, is much smaller than that used in water electrolysis.

Table 2.1 gives the values of the thermodynamic data under standard conditions calculated for the electrocatalytic oxidation of water and of several organic compounds with one to six carbon atoms.

Table 2.1 shows clearly that for most of these organic compounds, the theoretical total energy Δh^{+0} necessary to produce one mole of hydrogen by their electrochemical dissociation under standard conditions is at least

Table 2.1 Standard thermodynamic data, cell voltage, and number of hydrogen moles, N_{H_2}, produced by the electrochemical reforming of different hydrogen containing compounds.

Compound	N_{H_2}/mole	ΔH^{+0}/kJ mol^{-1}	Δh^{+0}/kJ mol H_2^{-1}	ΔG^{+0} /kJ mol^{-1}	$U^0_{cell} \approx E^+_a = +\Delta G^{+0}/nF$ (V vs SHE)
H_2O	1	+ 286	+ 286	+ 237	1.23
HCOOH	1	+ 32	+ 32	− 33	− 0.17
CH_3OH	3	+ 131.2	+ 44	+ 9.3	0.016
C_2H_5OH	6	+ 348	+ 58	+ 97.3	0.084
CH_2OH-CH_2OH	5	+ 240	+ 48	+ 17.2	0.018
1-C_3H_7OH	9	+ 545	+ 61	+ 170	0.098
DMM$CH_2(OCH_3)_2$	8	+ 340.6	+ 42.6	− 5.6	− 0.004
CH_2OH-CHOH-CH_2OH	7	+ 350	+ 50	− 1	− 0.001
1-C_4H_9OH	12	+ 754	+ 63	+ 205	0.177
Glucose	12	627.1	+ 52.3	− 27.4	− 0.012

five times lower than that for water electrolysis. Moreover, the theoretical electrical energy, which is proportional to the cell voltage—see Eq. (2.11) in Section 2.2—is at least several times lower than that for water dissociation, or even negative, for example, for HCOOH dissociation, which means that it is a spontaneous electrochemical process although the total energy ΔH^{+0} is positive due to a larger positive reversible heat transfer $T\Delta S$ associated with the entropy change.

2.2 Kinetics limitations

However, the kinetics of the anodic reactions involved in both processes, that is, water oxidation in the water electrolysis process or the electrocatalytic oxidation of organic compounds in the electrochemical reforming process, is relatively slow leading to high anodic overpotentials and thus to higher cell voltages at the relatively high current density, over 1 A cm^{-2}, required for a high hydrogen production rate (Fig. 2.2).

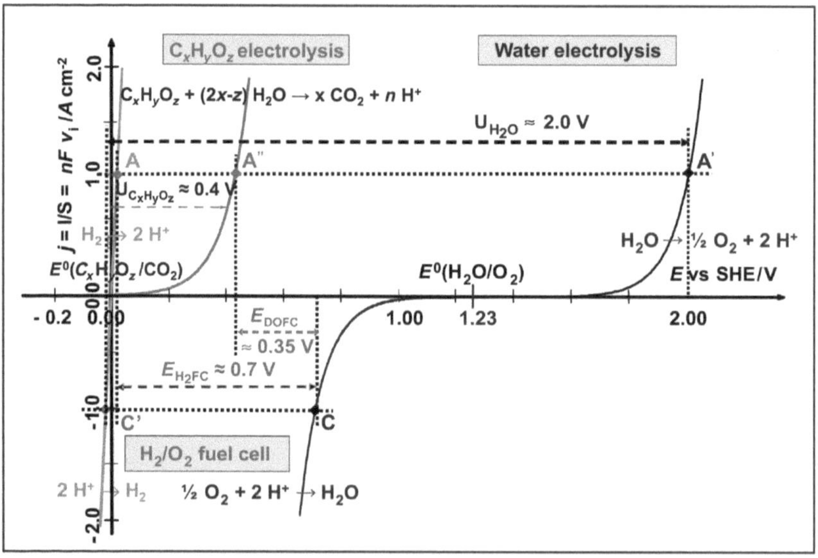

Figure 2.2 Theoretical electrical characteristics *j*(*E*) for a reaction kinetics controlled by the Butler – Volmer law: $C_xH_yO_z$ oxidation, H_2O oxidation, O_2 reduction, and proton reduction. $U_{C_xH_yO_z}$, U_{H_2O}, U_{DOFC}, and U_{H_2FC} are the cell voltages for $C_xH_yO_z$ electrolysis, water electrolysis, direct oxidation fuel cell, and hydrogen/oxygen fuel cell at a given current density *j*, for example, 1 A cm^{-2}.

This figure gives the current density $j = I/S = nFv/S = nFv_i$, where I is the current intensity, S is the electrode surface area, v and v_i are the reaction rate and the intrinsic rate, respectively, as a function of the electrode potential $E_i(j)$ versus the SHE taken as reference, for the different electrochemical processes involved:

- HOR and HER, which are described by a quasi-linear plot, assume that the hydrogen exchange current density j_{o_c} (of the order of 10^{-3} A cm^{-2}) is at least 10^3 times greater than the exchange current density j_{o_a} of the anodic processes. These processes are associated, either to the electrocatalytic oxidation of the organic compound or to the OER in water electrolysis, with $j_{o_a} \approx 10^{-6}$ A cm^{-2}, whereas the ORR has $j_{o_c} \approx 10^{-6}$ A cm^{-2}.
- In these latter cases, the current density j versus $E_i(j)$ curves are exponential functions when assuming no mass transfer limitation according to the Butler − Volmer law:

$$j(\eta) = j_o[\exp\{\alpha(nF/RT)\eta\} - \exp\{-(1-\alpha)(nF/RT)\eta\}] \quad (2.8)$$

where $\eta = E_i - E_{eq}$ is the overpotential, j_o is the exchange current density, and α is the anodic transfer coefficient.

- For irreversible reactions such as the OER/ORR or the electrochemical oxidation of an organic compound, we can neglect the opposite reaction, and the $j(E)$ characteristics become an exponential function, as plotted in Fig. 2.2, that is:

  $j(\eta) = j_o \exp\{\alpha(nF/RT)\eta\}$ for the oxidation of an organic compound or water in water electrolysis and

$$j(\eta) = -j_o \exp\{-(1-a)(nF/RT)\eta\} \quad \text{for the ORR}$$

When taking into account the charge transfer overpotentials η^{act}, the concentration overpotentials η^{conc} and the ohmic drop $R_e\, j$ associated with the cell resistance R_e, the cell voltage versus current density curves, $U_{cell}(j)$, of an electrolysis cell, can be expressed as follows:

$$\begin{aligned} U_{cell}(j) = E_a^+(j) - E_c^-(j) + R_e|j| = U_{cell}(0) + (|\eta_a^{act}(j)| \\ + \eta_a^{conc}(j)| + |\eta_c{}^{act}(j)| + |\eta_c^{conc}(j)|) + R_e|j| \end{aligned} \quad (2.9)$$

where $U_{cell}(0) = E_{eq}$ is the equilibrium cell voltage, that is, the cell voltage at zero current, η_i is the overpotential defined as the deviation of the electrode potential $E_i(j)$ under working conditions from its equilibrium value, that is, $\eta_i = E_i(j) - E_{i_{eq}}(0)$, and the subscripts "*a*" and "*c*" refer to the anodic and cathodic reactions, respectively.

For an electrochemical reaction not limited by mass transfer, that is, the corresponding concentration overpotentials $\eta^{conc} \rightarrow 0$, and assuming a high exchange current density j_{0c} for HER, equation (2.9) can be expressed as follows:

$$\begin{aligned} U_{cell}(j) &= U_{cell}(0) + \frac{RT}{nF}\frac{|j|}{|j_{0_c}|} + \frac{RT}{\alpha nF}\ln\frac{|j|}{j_{0_a}} + R_e|j| \\ &= U_{cell}(0) + R_{exp}|j| + \frac{RT}{\alpha nF}\ln\frac{|j|}{j_{0_a}} \end{aligned} \qquad (2.10)$$

so that $U_{cell}(j) - U_{cell}(0) - R_{exp}\ |j|$ can be plotted as a logarithmic function of the current density j, where $R_{exp} = R_e + RT/(nFj_{o_c})$ is the total contribution of the cell resistance, R_e, and the hydrogen transfer resistance $R_t = RT/(nFj_{o_c})$ to the linear part of the semi-logarithmic plot [6].

A specific software developed in our laboratory allowed us to estimate the unknown R_{exp}.

At a given current density, for example, $j = 1$ A cm^{-2} as in Fig. 2.2, the measured cell voltage, $U_{cell}(j)$, corrected from the ohmic losses $R_{exp}\, j$, allows to evaluate the electrical energy consumed to produce 1 Nm^3 of hydrogen, which is directly proportional to $U_{cell}(j)$, according to equation (2.11) [7]:

$$W_e\left(\text{in kWh/Nm}^3\right) = \frac{\tilde{n}F}{3600\,V_m \times 10^3}U_{cell}(j) \approx 2.191\,U_{cell}(j) \approx 2.2\ U_{cell}(j) \qquad (2.11)$$

where $\tilde{n} = 2$ is the number of Faraday per mole of hydrogen produced and $V_m = 24.465\ 10^{-3}$ m^3 mol^{-1} is the molar volume of an ideal gas at a temperature of 25°C under a pressure of 1 atm., that is, 101.325 kPa, corresponding to normal conditions for gaseous species.

Equation (2.11) shows that the electrical energy consumed to produce molecular hydrogen by the electro-oxidation of an organic compound is directly proportional to $U_{cell}(j)$, which must be as small as possible to decrease significantly the energy consumed, for example, below 1.0 kWh (Nm^3)$^{-1}$ for $U_{cell}(j) \leq 0.45$ V.

In contrast, the rate of hydrogen produced can be evaluated by measuring the volume of evolved hydrogen, V_{H_2}, which is a linear function of time for several fixed current intensities I during the electrolysis of the organic compound. The volume of evolved hydrogen, at a given electrolysis time Δt, depends only on the current intensity I, since it is a linear function of the quantity of electricity involved $\Delta q = I\ \Delta t$. The linear function can be derived from Faraday's law, equation (2.12a) or (2.12b), giving the rate of hydrogen evolution [7]:

$$\mathrm{d}V_{H_2}/\mathrm{d}t = V_{mol}(\mathrm{d}N_{H_2}/\mathrm{d}t) = V_{mol}\left(I/\tilde{n}\,F\right) \tag{2.12a}$$

$$\mathrm{d}V_{H_2}/\mathrm{d}t = V_{mol}(I/2\mathrm{F}) \times 60 = 7.607\ I(\text{in cm}^3\ \text{min}^{-1}) \tag{2.12b}$$

where N_{H_2} is the number of hydrogen mole, $F = 96{,}485$ C per mole of electron is the Faraday constant, I is the current intensity in Ampere, and $\tilde{n} = 2$ is the average number of Faraday per mole of dihydrogen produced in the complete oxidation reaction of an organic molecule according to reaction (2.4) or (2.6).

The volume of hydrogen produced after $\Delta t = 15$ min of electrolysis at a given temperature, when a quasi-stationary state is reached, can be obtained by integration of equation (2.12a), that is:

$$\begin{aligned} V_{H_2} &= V_{mol}(I\ \Delta t/2F) = 24.465\ 10^{-3}\ I(60 \times 15)/(2 \times 96485) \\ &= 114.1 \times I(\text{in cm}^3) \quad \text{at } 25^\circ\text{C} \end{aligned} \tag{2.13a}$$

$$\text{or } V_{H_2} = 119.8 \times I\ \text{cm}^3 \text{ at } 40^\circ\text{C} \quad \text{and} \quad V_{H_2} = 131.3 \times I\ \text{cm}^3 \quad \text{at } 70^\circ\text{C} \tag{2.13b}$$

when taking into account the hydrogen molar volume at 40°C or 70°C, corresponding to the dilatation of an ideal gas. This confirms that V_{H_2} is proportional to the current intensity I leading to linear plots $V_{H_2}(I)$.

2.3 Analysis of the cell voltage versus current characteristics

When analyzing the cell voltage, $U_{cell}(j)$, recorded at a given current density j under several experimental conditions of temperature T and pressure p, equation (2.10) can be written, after correction of ohmic drop $R_e\ |j|$, as follows:

$$U_{cell}(j) - U_{cell}(0) - R_{exp}|j| = \frac{RT}{\alpha nF}\ln\frac{|j|}{j_{0_a}}$$

where $R_{exp} = R_e + RT/(nFj_{o_c})$ is the total contribution of the cell resistance, R_e, and the hydrogen transfer resistance $R_t = RT/(nFj_{o_c})$ to the linear part of the semi-logarithmic plot.

The effect of temperature T, at the standard pressure p_0, is contained in $U_{cell}(0) = U_{rev}(0) = E_{eq}(0)$, that is, in the Gibbs energy change with T at $j = 0$:

$U_{cell}(0) = E_a^+ - E_c^- \approx E_a^+ = (\Delta G^+ - \Delta G^-)/nF \approx \Delta G^+(T)/nF$ since $\Delta G^- = 0$ at any temperature for the hydrogen electrode, with $n = (4x + y - 2z)$ is the number of Faraday involved in the complete oxidation of $C_xH_yO_z$, into H_2 and CO_2—see eq. (2.6). For a given organic compound, $\Delta G^+(T)$ can be evaluated from the variation of ΔG_{H_2O}, ΔG_{CO_2}, and $\Delta G_{C_xH_yO_z}$ with T according to the following relation:

$$\Delta G^+(T) = x\,\Delta G_{CO_2}(T) - (2x - z)\,\Delta G_{H_2O}(T) - \Delta G_{C_xH_yO_z}(T)$$

As an example, the complete calculation can be made for the generation of H_2 from methanol electro-oxidation:

$$CH_3OH + H_2O \rightarrow CO_2 + 3\,H_2$$

since the variations of ΔH and ΔG with T are available from 298.15 to 400 K (Table 2.2 and Fig. 2.3).

The variation of $U_{cell}(0)$ with T may be calculated from U_{tn} by the following differential thermodynamic law:

$$\frac{d}{dT}\left(\frac{\Delta G}{RT}\right) = -\frac{\Delta H}{RT^2} \quad \text{or} \quad \frac{d}{dT}\left(\frac{U_{cell}(0)}{T}\right) = -\frac{U_{tn}}{T^2}$$

where $U_{tn} = \Delta H/nF$ is the thermoneutral voltage with $\Delta H = 130.58 + 0.0349\ T{-}0.0001\ T^2 + \text{etc.} = a + b\ T + c\ T^2 + d\ T^3 + e\ T^4 + \text{etc.}$ One obtains

$$\frac{\Delta G}{T} = \frac{\Delta G^0}{T_0} - \int_{T_0}^{T} \frac{\Delta H}{T^2}\,dT = \frac{\Delta G^0}{T_0} + a\left(\frac{1}{T} - \frac{1}{T_0}\right)$$
$$- b\ \mathrm{Ln}\frac{T}{T_0} - c - d(T - T_0) + \text{etc.}$$

and

$$\Delta G(T) = \Delta G^0 \frac{T}{T_0} + a\left(1 - \frac{T}{T_0}\right) - b\ T\ \mathrm{Ln}\frac{T}{T_0} - c\ T$$
$$- d\ T(T - T_0) + \text{etc.}$$

Table 2.2 Thermodynamic data and open circuit cell voltage corresponding to the electrochemical reforming of methanol as a function of temperature between 298.15 and 400 K.

T (K)	ΔS (J mol K)$^{-1}$	ΔH (kJ mol^{-1})	ΔG (kJ mol^{-1})	$U_{cell}(0)$ (mV)
298.15	408.252	131.16	9.44	16.30
300.00	408.048	131.10	8.68	14.99
305.00	407.495	130.93	6.64	11.47
310.00	406.940	130.76	4.61	7.96
315.00	406.381	130.58	2.57	4.44
320.00	405.818	130.40	0.54	0.94
325.00	405.250	130.22	− 1.49	− 2.57
330.00	404.676	130.03	− 3.51	− 6.06
335.00	404.097	129.84	− 5.53	− 9.55
340.00	403.512	129.64	− 7.55	− 13.04
345.00	402.920	129.44	− 9.57	− 16.52
350.00	402.321	129.23	− 11.58	− 20.00
355.00	401.715	129.02	− 13.59	− 23.47
360.00	401.100	128.80	− 15.60	− 26.94
365.00	400.477	128.57	− 17.60	− 30.40
370.00	399.846	128.34	− 19.60	− 33.86
375.00	399.206	128.10	− 21.60	− 37.31
380.00	398.557	127.86	− 23.59	− 40.75
385.00	397.898	127.61	− 25.59	− 44.19
390.00	397.231	127.35	− 27.57	− 47.62
395.00	396.554	127.08	− 29.56	− 51.05
400.00	395.868	126.81	− 31.54	− 54.47

Another way to get the temperature variation of $U_{cell}(0)$ is to consider the entropic term, that is, $\Delta S_r^0 = nF\ (\mathrm{d}U_{cell}(0)/\mathrm{d}T)$, which is related to the temperature coefficient of the cell. Thus, knowing the variation of ΔH and ΔS with the temperature, one may calculate $U_{cell}(0)$ as follows:

$$U_{cell}(0) = \Delta G/nF = (\Delta H - T\Delta S)/nF$$

for water and methanol in the liquid state.

The effect of pressure for gases or concentration for liquids can be evaluated with the Nenst'law giving the electrode potential for each compartment:

$$U_{rev}(T_0, p) = E_{eq} = E_{eq}(T_0, p_0) + \frac{RT}{nF} \ln\left[\left(\frac{p_{CO_2}}{p_0}\right)\left(\frac{p_{H_2}}{p_0}\right)^3 / \left(\frac{[CH_3OH]}{[CH_3OH]_0}\frac{[H_2O]}{[H_2O]_0}\right)\right]$$

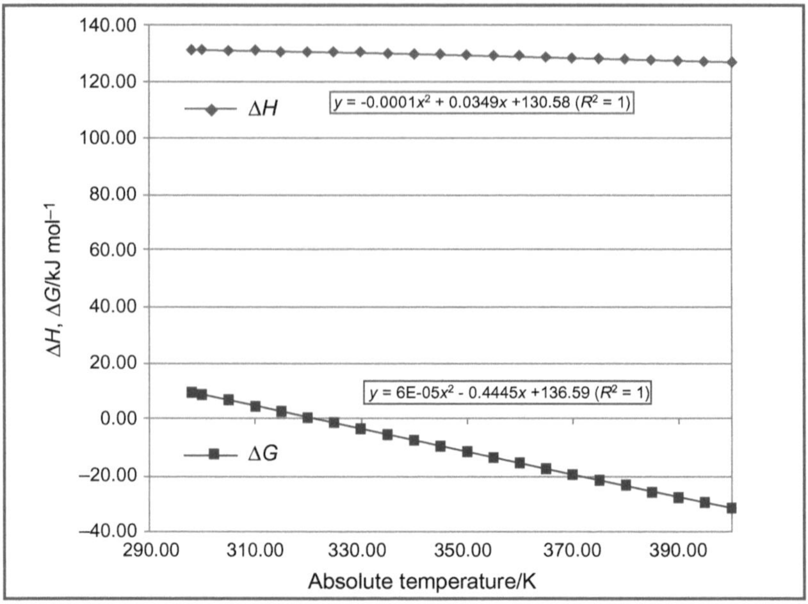

Figure 2.3 Enthalpy change ΔH^{+} and Gibbs energy change ΔG^{+} for the electrochemical reforming of methanol as a function of temperature between 298.15 and 400 K.

where U_{rev} $(T_0, p_0) = E_{eq}$ $(T_0, p_0) = \Delta G^0/nF$ is the equilibrium cell voltage in the standard state and [species (i)] the concentration of species (i) in the liquid state.

2.4 Estimation of the energy efficiency [8]

The energy efficiency ε_{cell} of an electrolysis cell can be defined as the ratio between the theoretical amount of total energy W_t (J mol^{-1}) required to decompose the organic compound $C_xH_yO_z$ with the production of $(2x + y/2 - z)$ mole of hydrogen, according to the overall equation (2.6), that is, the reaction enthalpy $\Delta H^0_{rev} = x\,\Delta H^0_{CO2} - \Delta H^0_f - (2x - z)\Delta H^0_{H2O}$, where ΔH^0_f are the formation enthalpy, under standard state, of the species involved and the real amount of energy W_r (J mol^{-1}) used in the process:

$$\varepsilon_{cell} = \frac{\text{energy requirement under reversible conditions}}{\text{energy requirement under irreversible conditions}} = \frac{W_t(\text{J mol}^{-1})}{W_r(\text{J mol}^{-1})} \tag{2.14}$$

When taking into account all the energies involved, that is, electrical and thermal energies, the numerator of Eq. (2.14), that is, the energy requirement under reversible conditions ($j = 0$), is defined as the necessary electrical work + the necessary heat flow $\Delta Q_{rev} = T\ \Delta S$ (in J mol^{-1}), associated with the entropy change. Therefore:

$W_t = \Delta G_{rev} + \Delta Q_{rev} = \Delta H_{rev} = nF\ U_{rev}$ (electrical work) $+ nF\ (U_{tn} - U_{rev})$ (reversible Q)

$W_t = nF\ U_{tn}$ (J mol^{-1}) (total energy) with U_{tn} the thermoneutral cell voltage ($U_{tn} = \Delta H_{rev}/nF$).

The denominator of Eq. (2.14), that is, the energy requirement under irreversible conditions ($j \neq 0$), is defined as the real electrical energy consumption, that is, the necessary electrical work + the extra amount of electrical work, which is dissipated internally into heat + the necessary heat associated with the entropy increase. Therefore:

$W_r = \Delta G_{rev} + nF\ \eta_{loss} + \Delta Q_{rev} = \Delta H_{rev} + nF\ \eta_{loss}$ $= nF\ U_{rev}$ (electrical work) $+ nF\ (U_{cell} - U_{rev})$ (irreversible Q) $+ nF\ (U_{tn} - U_{rev})$ (reversible Q), where the irreversible heat, defined as:

$$Q_{irrev}(\text{in J mol}^{-1}) = nF(U_{cell} - U_{rev}) = nF\ \eta_{loss} = nF\left(\sum |\eta_i| + R_e I\right)$$

represents the energy loss due to internal dissipation via the charge transfer overpotentials ($\Sigma|\eta_i|$) and the ohmic dissipation ($R_e\ I$).

Thus $W_r = nF\ (U_{tn} + U_{cell} - U_{rev})$ (total energy in J mol^1), and the "energy efficiency" is given by:

$$\varepsilon_{cell} = \frac{\Delta H_{rev}(j=0)}{\Delta H(j \neq 0)} = \frac{\Delta H_{rev}}{\Delta H_{rev} + nF\eta_{loss}} = \frac{\Delta G_{rev} + \Delta Q_{rev}}{nF\ U_{cell} + \Delta Q_{rev}} = \frac{U_{tn}}{U_{tn} + U_{cell} - U_{rev}} \qquad (2.15a)$$

If the reversible heat, $\Delta Q_{rev} = T\ \Delta S$, is small compared to ΔG_{rev}, that is, $\Delta Q_{rev}/\Delta G_{rev} << 1$, this expression can be simplified as follows:

$\varepsilon_{cell} \approx U_{rev}/U_{cell}$, which represents the *voltage efficiency* (2.15b)

However, for most decomposition reactions of organic compounds, the reversible heat, ΔQ_{rev}, is much greater than ΔG_{rev} so that only the full equation (2.15a) does apply.

As an example [7], the energy efficiency of an electrolysis cell fed with methanol at 70°C working at $I = 500$ mA and at a corrected cell voltage $U_{cell} = 0.488$ V leading to an energy loss is as follows:

$$nF\,\eta_{loss} = nF(U_{cell} - U_{rev}) = 6 \times 96.485(0.488 - (-0.015)) = 291.2 \text{ kJ}$$

and a decomposition reaction enthalpy $\Delta H_{rev} = 129.5$ kJ (see Table 2.2) would be:

$$\varepsilon_{cell} = 129.5/(129.5 + 291.2) = 0.308 \approx 31\%$$

If instead of using the complete formula (2.15a), the simplified expression (2.15b) would give:

$$\varepsilon_{cell} = 0.015/0.488 = 0.0307 \approx 3.1\%, \quad \text{which is completely wrong}$$

References

[1] Y.H. Hu, E. Ruckenstein, Binary MgO-based solid solution catalysts for methane conversion to syngas, Catal. Rev. Sci. Eng. 44 (2002) 423–453. Available from: https://doi.org/10.1081/CR-120005742.
[2] F. Auprêtre, C. Delorme, D. Duprez, Bio-ethanol catalytic steam reforming over supported metal catalysts, Catal. Commun. 3 (2002) 263–267. Available from: https://doi.org/10.1016/S1566-7367(02)00118-8.
[3] G.A. Deluga, J.R. Salge, L.D. Schmidt, X.E. Verykios, Renewable hydrogen from ethanol by autothermal reforming, Science 303 (2004) 993–997. Available from: https://doi.org/10.1126/science.1093045.
[4] D.K. Liguras, K. Goundani, X.E. Verykios, Production of hydrogen for fuel cells by catalytic partial oxidation of ethanol over structured Ru catalysts, Int. J. Hydrogen Energy 29 (2004) 419–427. Available from: https://doi.org/10.1016/S0360-3199(03)00210-6.
[5] C. Lamy, Operation of fuel cells with biomass resources (hydrogen and alcohols), in: P. Lens, C. Kennes, P. Le Cloirec, M. Deshusses (Eds.), Waste gas treatment for resource recovery, IWA Publishing, London, 2006, pp. 360–384. ISBN13: 9781843391272.
[6] C. Lamy, B. Guenot, M. Cretin, G. Pourcelly, Kinetics analysis of the electrocatalytic oxidation of methanol inside a DMFC working as a PEM electrolysis cell (PEMEC) to generate clean hydrogen, Electrochim. Acta 177 (2015) 352–358. Available from: https://doi.org/10.1016/j.electacta.2015.02.069.
[7] B. Guenot, M. Cretin, C. Lamy, Clean hydrogen generation from the electrocatalytic oxidation of methanol inside a proton exchange membrane electrolysis cell (PEMEC): effect of methanol concentration and working temperature, J. Appl. Electrochem. 45 (2015) 973–981. Available from: https://doi.org/10.1007/s10800-015-0867-3.
[8] C. Lamy, From hydrogen production by water electrolysis to its utilization in a PEM fuel cell or in a solid oxide fuel cell: some considerations on the energy efficiencies, Int. J. Hydrogen Energy 41 (2016) 15415–15425. Available from: https://doi.org/10.1016/j.ijhydene.2016.04.173.

CHAPTER THREE

Experimental part

Contents

3.1 Preparation and physicochemical characterization of the electrocatalysts

3.1.1 Preparation of catalysts by the water-in-oil microemulsion method

The experiments on formic acid, glycerol, and glucose were performed using catalysts synthesized by a water-in-oil microemulsion method [1,2]. Appropriate amounts of metal salts were first dissolved in 1.6 mL ultrapure water (MilliQ, Millipore, 18.2 MΩ cm) to reach the desired atomic ratios with a total metal concentration of 0.2 mol L^{-1}. Hexachloroplatinic acid hexahydrate (H_2PtCl_6•$6H_2O$), tetrachloroauric acid trihydrate ($HAuCl_4$•$3H_2O$), potassium tertrachloropalladate (K_2PdCl_4), nickel chloride ($NiCl_2$), tin chloride ($SnCl_2$) and bismuth chloride ($BiCl_3$) 99% purity from Alfa Aesar were used. In a second flask, the oil phase was prepared by mixing 37.0 g of n-heptane (Sigma–Aldrich, high-performance liquid chromatography [HPLC] grade) and 16.1 g of polyethylene glycol dodecyl ether (Brij L4, Sigma–Aldrich) under continuous stirring conditions until a homogeneous solution was obtained. The aqueous phase

Production of Clean Hydrogen by Electrochemical Reforming of Oxygenated Organic Compounds.
DOI: https://doi.org/10.1016/B978-0-12-821500-5.00003-5

containing the metal salts was added to the oil phase containing the Brij L4 surfactant under continuous stirring until a translucent and stable microemulsion was obtained. Solid sodium borohydride ($NaBH_4$, 99% purity, Sigma–Aldrich) was added in large excess, c.100–200 mg, to the mixture to reduce the metal salts. Carbon, such as Vulcan XC72 CABOT, pretreated under $N_2(g)$ at 400°C for 4 hours, was directly added to the colloidal solution in order to reach a metal loading of c.40 wt.%. The suspension was ultra-sonicated for c.15 minutes to facilitate the dispersion of the metal nanoparticles on the carbon substrate. The mixture containing the carbon-supported nanoparticles was then filtered using a hydrophilic polyvinylidene difluoride (PVDF) membrane such as a 0.22 mm Durapore membrane filter from Millipore. The carbon-supported metal nanoparticles were washed several times with acetone and ultrapure water and dried overnight in an oven under air atmosphere at 60°C. Finally, the catalytic powder was thermally treated for 2 hours at 200°C under air atmosphere to remove any remaining surfactant.

3.1.2 Preparation of catalysts by the Bönnemann method

The experiments on ethanol electrooxidation were performed on catalysts prepared by the Bönnemann method [3,4]. The synthesis was carried out under an argon controlled atmosphere free of oxygen and water, with non-hydrated metal salts (99.9% $PtCl_2$, $RuCl_3$, and $SnCl_2$ from Alfa Aesar). The tetraalkyl triethylborohydride reducing agent, $(Nalk_4)^+(BEt_3H)^-$, was first synthesized. This reactant will also act as a surfactant after metal reduction, preventing any particle agglomeration after the formation of colloidal particles, according to the following reaction:

$$MCl_n + n\,(Nalk_4)^+(BEt_3H)^- \rightarrow M\left[(Nalk_4)^+Cl^-\right]_n + n\,BEt_3 + (n/2)\,H_2 \tag{3.1}$$

The colloidal precursors are then dispersed on a carbon support, such as Vulcan XC72, and calcined at 300°C for 1 hour under air atmosphere to remove the organic surfactant. The synthesis of $Pt_{0.9}Sn_{0.1}/C$ and $Pt_{0.86}Sn_{0.1}Ru_{0.04}/C$ (atomic ratios) catalysts was carried out by coreduction of the different metal salts during the formation of the colloidal precursor according to the following equation:

$$x\,PtCl_2 + y\,MCl_n + (2x+ny)\left[(Nalk_4)^+(BEt_3H)^-\right]$$
$$\rightarrow Pt_xM_y\left[(Nalk_4)^+Cl^-\right]_{(2x+ny)} + (2x+ny)\,BEt_3 + \left(x+ny/2\right)H_2 \quad (3.2)$$

3.1.3 Physicochemical characterization of catalysts

The synthesized catalysts were characterized by thermogravimetric analysis (TGA) measurements, transmission electron microscopy (TEM), X-ray diffraction (XRD), energy dispersive X-ray spectrometry (EDS), inductively coupled plasma optical emission spectroscopy (ICP-OES), and atomic absorption spectroscopy (AAS). The metal loadings were determined by TGA analysis carried out using a TA Instrument SDT Q600 apparatus. The catalytic powder (c.5 mg) was put in an alumina crucible and heated under air from 25°C to 900°C with a temperature slope of 5°C $minute^{-1}$. AAS analyses were carried out with a Perkin Elmer AA200 Atomic absorption spectrometer. TEM images were acquired using either a Philips CM 120 microscope (120 kV) or a JEOL JEM 2100 (UHR) with a resolution of 0.19 nm. The mean particle sizes and nanoparticle size distributions were determined from the Feret's diameters by counting at least 200 isolated nanoparticles using the ImageJ software [5]. X-ray powder diffraction patterns were recorded either on a Bruker D5005 Bragg Brentano ($\theta-\theta$) diffractometer or on a PANalytical "Empyrean" diffractometer in step mode (0.06° step and acquisition time of 10 seconds/step). Results are given in Table 3.1.

The metallic composition and electronic state of the catalysts can be analyzed by X-ray photoelectron spectroscopy (XPS) with a Kratos Axis Ultra DLD spectrometer. An Al anode was used as source operating ($K_\alpha = 1486.6$ eV). The pressure in the analytical chamber of the spectrometer was kept below 8×10^{-9} Torr. For each sample analyzed, a survey scan was performed, followed by a core level spectrum of the different metals composing the catalysts. The binding energies were lined up with respect to the C 1-second peak at 284.6 eV.

As a typical example, XPS spectra recorded from 0 to 800 eV for a PtSn catalyst before and after a long duration of electrolysis experiments to investigate the oxidation state of Sn, which are given in Fig. 3.1 [6]. The XPS spectra for Pt and Sn showed that there is no shift of binding energy of Pt $4f_{5/2}$ (74.5 eV) and $4f_{7/2}$ (71.3 eV) peaks of the PtSn catalyst before and after a long duration electrolysis as seen in Fig. 3.1(b).

Table 3.1 Main characterization data for some of the catalysts used in this book chapter, determined from TGA[(1)], XRD[(2)], TEM[(3)], and EDX[(4)] for the catalysts prepared by the Bönnemann method and ICP-OES for the catalysts prepared by the water-in-oil microemulsion method, except for $Pt_{0.3}Pd_{0.6}Bi_{0.1}/C$, which was analyzed by AAS[(5)].

Synthesis method	Metal loading (1) (wt.%)	Crystallite size (2) (nm)	Particle size (3) (nm)	Atomic composition (4) (%)			
Bönnemann				*%Pt*	*%Pd*	*%Au*	*%M*
Pt/C		2.1	2.4			100	
$Pt_{0.9}Sn_{0.1}/C$		1.9	2.4	90.2			9.8
$Pt_{0.86}Ru_{0.04}Sn_{0.1}/C$		–	–			–	
Water-in-oil							
Pt/C	37	5.5	5.3	100			
Pd/C	36	2.9	4.0	–	100	–	–
Au/C	39	8.0	7.4	–	–	100	–
$Pt_{0.5}Pd_{0.5}/C$	–	–	–				
$Pt_{0.9}Bi_{0.1}/C$	36	5.7	4.7	90			10
$Pd_{0.5}Ni_{0.5}/C$	43	3.4	5.5		54		46
$Pd_{0.9}Bi_{0.1}/C$	38	4.5	5.2		92		8
$Pd_{0.9}Au_{0.1}/C$	35	3.7	4.7		90	10	
$Pd_{0.7}Au_{0.3}/C$	37	3.7	4.9		69	31	
$Pd_{0.5}Au_{0.5}/C$	37	3.9	5.0		50	50	
$Pd_{0.3}Au_{0.7}/C$	40	4.6	5.1		31	69	
$Pd_{0.1}Au_{0.9}/C$	38	5.4	5.3		11	89	
$Pt_{0.45}Pd_{0.45}Bi_{0.1}/C$	38	4.7	4.5	45	43		12
$Pt_{0.3}Pd_{0.6}Bi_{0.1}/C$[(5)]	34	3.3	–	26	64		10

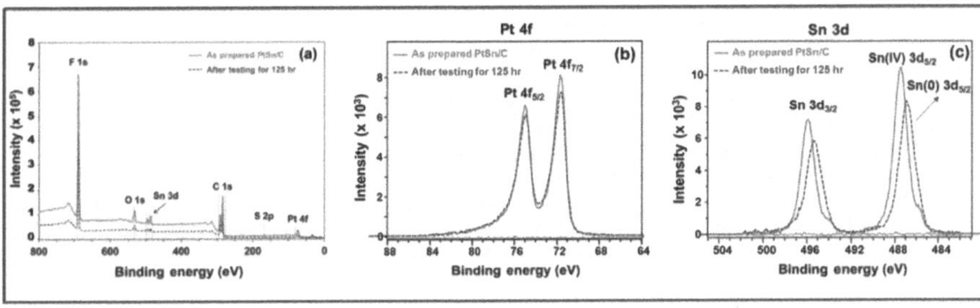

Figure 3.1 XPS spectra of the catalyst before and after testing a PtSn/C anode for around 125 h. (a) XPS wide-scan spectra, (b) Pt 4f spectrum, and (c) Sn 3d spectrum. *After [6].*

However, Fig. 3.1(c) displayed that the binding energy of Sn $3d_{5/2}$ peak shifted to a lower value. The Sn $3d_{5/2}$ peak located at 487.5 eV was assigned to oxidized Sn in the as-prepared catalyst, while the $3d_{5/2}$ peak observed at 486.9 eV for the tested anode appeared between the peak for metallic Sn and SnO_2. The SnO_2 catalyst surface of the anode is gradually reduced to a lower oxidation state (SnO_x) during electrolysis [6].

3.1.4 Electrochemical characterization of catalysts

Polarization curves were recorded at 20°C in a classical three-electrode electrochemical cell using a Voltalab PGZ 402 computer-controlled potentiostat. The reference electrode was a reversible hydrogen electrode (RHE) and the counter electrode was a 3 cm^2 glassy carbon plate. The working electrode was prepared from a catalytic ink containing the catalytic powder (25 mg) suspended in a 0.5 mL Nafion solution (5 wt.% from Aldrich) and in 2.5 mL ultrapure water. After ultrasonic homogenization of the catalyst/C-Nafion ink, 3 μL of the ink were dispersed with a microsyringe onto a freshly polished glassy carbon substrate, used as a working electrode of 0.071 cm^2 geometric surface area, yielding a catalytic powder with loading of c.354 $\mu g\ cm^{-2}$, that is, a metal loading of c.141 $\mu g_{metal}\ cm^{-2}$ for catalysts with 40 wt.% metal on carbon. The solvent was then evaporated in a stream of ultrapure nitrogen at room temperature.

The electrooxidation of the organic compounds was carried out either in nitrogen-purged (N_2 U-quality, Air Liquide) acidic or alkaline supporting electrolytes. The acidic electrolyte was 0.5 M H_2SO_4 or 0.1 M $HClO_4$ (Suprapur, Sigma—Aldrich) in ultrapure water. The alkaline solution was 0.1 or 1.0 M NaOH (Semiconductor Grade 99.99%, Sigma—Aldrich) dissolved in ultrapure water. Formic acid (97%, Alfa

Aesar), methanol (Chromatosolv for HPLC ≥ 99.9%, Sigma–Aldrich), ethanol (absolute for HPLC ≥ 99.8%), glycerol (Reagent Plus ≥ 99%, Sigma–Aldrich), glucose and xylose (99%, Sigma–Aldrich) were used as reactants.

To determine the potential range where intermediate and reaction products are formed, *in* situ single potential alteration infrared spectra [7] were recorded by using a Bruker IFS 66v FTIR spectrometer modified for beam reflection on the electrode surface at a 65 degree incident angle. To remove interferences from atmospheric water and carbon dioxide, the beam path was evacuated. An Infrared Associates liquid-nitrogen-cooled HgCdTe detector was used. The spectral resolution was 4 cm^{-1}, and each spectrum was obtained by averaging 256 spectra recorded for 35 seconds. The spectra were recorded every 0.05 V during the linear sweep voltammetry performed at a scan rate of 1 mV $second^{-1}$ so that each spectrum represents the average signal recorded in a 0.035 V range every 0.05 V. The spectra were normalized as the relative change of reflectivity $\Delta R/R$, that is, the opposite of the absorbance change ΔA due to the presence of IR absorbing species at the electrode surface, according to the following relation:

$$-\Delta A = \frac{\Delta R}{R} = \frac{R_E - R_{E_{\mathrm{Ref}}}}{R_{E_{\mathrm{Ref}}}}$$

where R_E is the reflectance spectrum recorded at a potential E during the positive potential scan and $R_{E\mathrm{Ref}}$ is that recorded at a reference potential of 0.1 V versus RHE, where no IR absorbing species are present at the electrode surface. Therefore, negative absorption bands correspond to the formation of species at the electrode surface and positive absorption bands correspond to the consumption of species at the electrode surface.

3.2 Fabrication of the membrane-electrode assembly

3.2.1 Investigation in acidic medium

The electrolysis cell tests in a single proton exchange membrane electrolysis cell (PEMEC) with a 5 cm^2 geometric surface area were carried out by feeding the anodic compartment with a formic acid (1–10 M) or methanol (0.1–10 M) or ethanol (2 M) aqueous solution and the cathodic compartment with an acidic solution (0.5 M H_2SO_4) at a flow rate of 2 mL minute^{-1}. In the case of formic acid, the $U_{cell}(t)$ and $U_{cell}(j)$ curves were recorded using a high power potentiostat (Wenking model LT 87, Bank Electronik) and a variable resistance in order to fix the current applied to the cell, whereas in the case of ethanol, the $U_{cell}(t)$ and $U_{cell}(j)$ curves were recorded using a DC Power Supply (E3614A from Agilent) to fix the current and Digital Multimeters (34405 A from Agilent) to record the cell voltage and the applied current density. The volume of the generated H_2 was measured by a gas volumeter consisting in the water displacement in a graduated glass tube connected to the cell. The experimental setup is schematized in Fig. 3.6 (see Section 3.3).

Carbon porous layers (CPLs) were home-made using a carbon cloth from Electrochem Inc. on which was brushed an ink made of Vulcan XC72 carbon powder and PTFE dissolved in isopropanol. The CPLs were loaded with 4 mg cm^{-2} of a mixture of carbon powder and 20 wt.% PTFE. The catalytic electrodes for PEMEC were prepared from an ink consisting in a mixture of Nafion (5 wt.% from Aldrich) solution, ultrapure water and the catalytic powder, dried on a CPL. The metal loading of the catalytic electrodes was close to 2.0 mg cm^{-2} and the Nafion loading of the electrode was c.0.8 mg cm^{-2}. The membrane-electrode assembly (MEA) was prepared by hot pressing at 130°C for 90 seconds under a pressure of 35 kg cm^{-2}, a pretreated Nafion 117 membrane with a cathode (2.0 mg cm^{-2} Pt loading, 20 wt.% PTFE, 0.8 mg cm^{-2} Nafion), and an anode (2.0 mg cm^{-2} metal loading, 20 wt.% PTFE, 0.8 mg cm^{-2} Nafion). The metal loadings on the carbon powder for both electrodes (anode and cathode) were 30 wt.% metal for the experiments with ethanol and 40 wt.% metal for the experiments with formic acid.

3.2.2 Investigation in alkaline medium

The electrolysis cell tests were carried out in a recirculation mode at 20°C with 2 M glycerol and × M NaOH aqueous solutions in a single electrolysis cell with a 5 cm^2 geometric surface area or with 0.1 mol L^{-1} glucose and 0.1 mol L^{-1} NaOH aqueous solution in a single electrolysis cell with a 25 cm^2 geometric surface area, at a flow rate of 20 mL minute^{-1}. The $U_{cell}(t)$ curves were recorded by using a DC power supply (E3614A from Agilent) to fix the cell voltage and digital multimeters (34405 A from Agilent) to record the cell voltage and the applied current.

The same CPLs were used for acidic environment. In the case of glycerol, the metal loading of both electrodes (anode and cathode) was 1.6 mg cm^{-2} and the Nafion loading of the electrode was c.0.8 mg cm^{-2}. In the case of glucose, the metal loading of both electrodes (anode and cathode) was 0.5 mg cm^{-2} and the Nafion loading of the electrode was 0.125 mg cm^{-2}. The electrodes (cathode and anodes) were separated by a simple blotting paper and mechanically pressed in the cell. The experimental setup for the measurement of hydrogen produced is the same as for acidic medium.

3.3 Electrolysis experiments

The electrooxidation of an organic compound for producing clean hydrogen can be carried out in a PEMEC with different setup configurations and experimental conditions of the experimental system, as described in the literature: [8–11]

a. In the earlier experiments using methanol as hydrogen source [8,10] the electrolysis cell consisted of an MEA inserted in a configuration similar to that encountered in a fuel cell (Fig. 3.2).

Pt-based catalysts are usually chosen for both electrodes and a Nafion 117 membrane is often used as the protonic membrane. The electrolysis can be carried out either at a constant current intensity or a constant cell voltage using a DC power supply. The first protocol is more convenient since the flow rate of hydrogen evolution is proportional to the current intensity (see Eq. (12) in Section 2.2). An entire experimental setup is shown in Fig. 3.3 [8].

Both liquid solutions were supplied to their respective compartments at a controlled constant flow rate between 5 and 12 mL $minute^{-1}$. The cathode exhaust gas was analyzed by gas chromatography (GC) and measured by a flow meter. A thermometer was used to measure the cell temperature. Both the cell voltage and the current intensity were measured with a digital voltmeter and an ammeter respectively.

b. A second approach is to insert the MEA between the two compartments of a dual-chamber glass cell as shown in Fig. 3.4 [11].

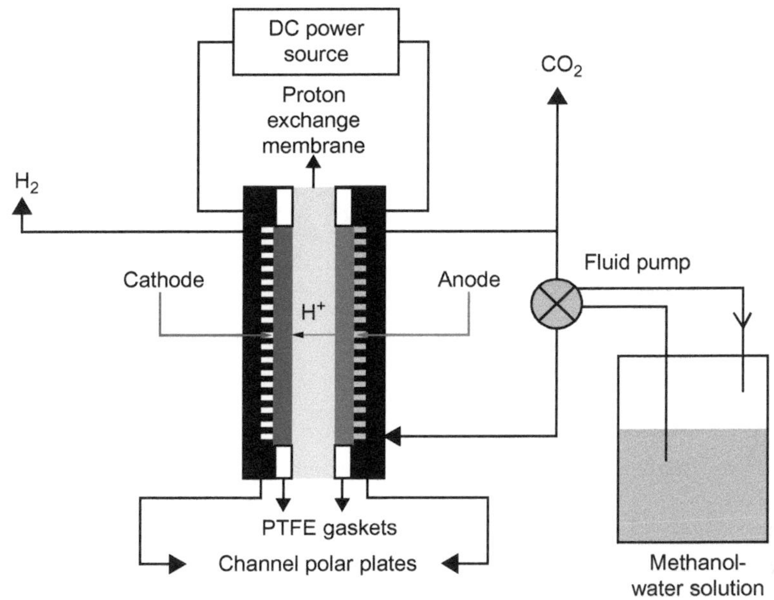

Figure 3.2 Schematic representation of the electrolyzer used to produce hydrogen by methanol electrolysis. *Modified after [10].*

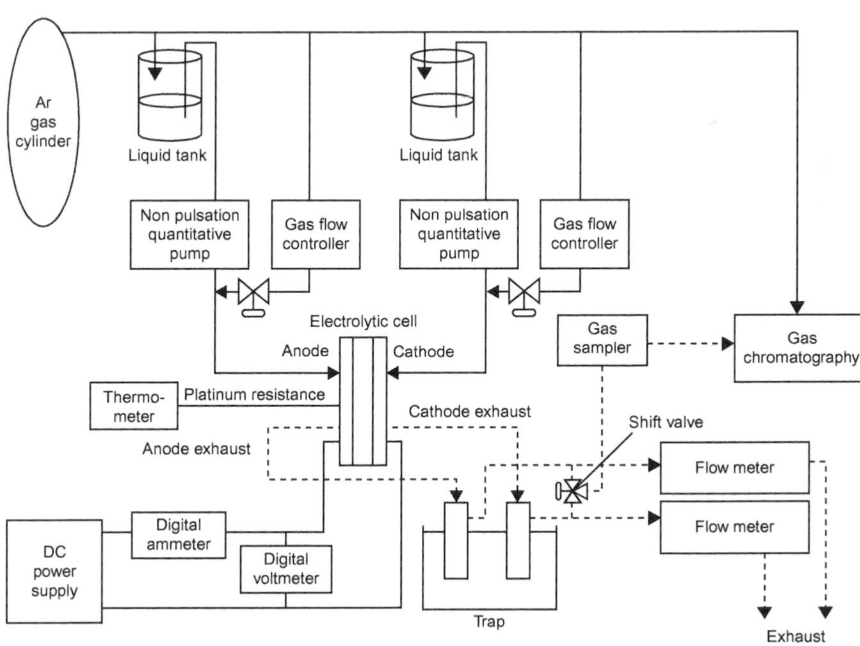

Figure 3.3 Schematic diagram of an experimental apparatus to investigate hydrogen production by the electrolysis of methanol-water solutions. *After [8].*

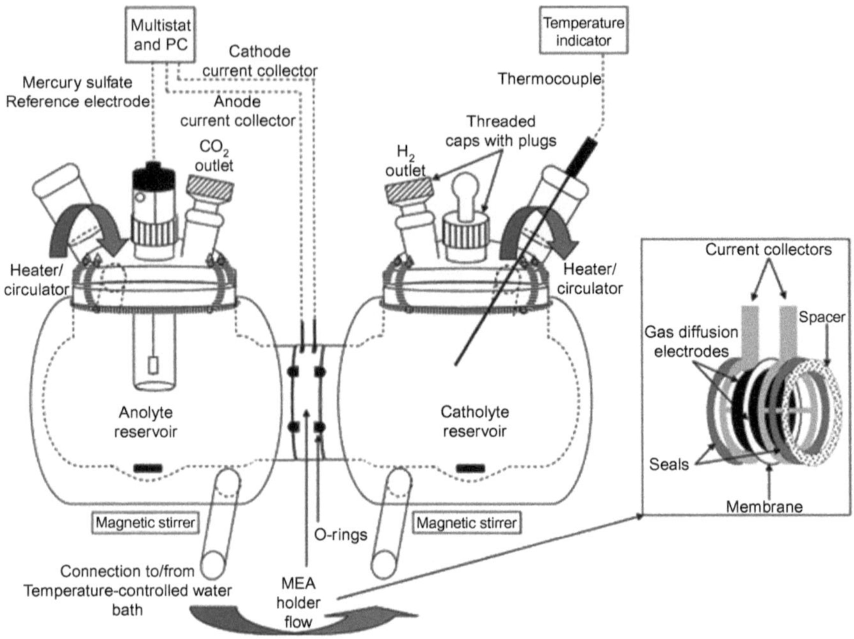

Figure 3.4 Schematic of the experimental apparatus for methanol electrolysis. *After [11].*

Various ports allow the inclusion of reference electrodes, thermocouples, gas evolution-inverted 10-mL burettes, and gas bubblers in the anode or cathode compartments. Slight internal bevels were provided in the upper part of the anode and cathode compartments to facilitate the gas exit. The CO_2 gas forming at the anode and H_2 gas forming at the cathode were vented in all experiments. The MEA holder has a central opening with a circular active geometric area of 2 cm^2. The MEA holder contained different layers: the MEA, the current collectors, two silicon sealing rings, and a PTFE spacer ring. A 25 mm diameter membrane coupon is sandwiched between the anodic and cathodic 16 mm diameter gas diffusion electrode (GDE) layers. Two niobium current collectors (25 mm OD, 16 mm ID), having a thin current collecting metal cross through their circular openings, are placed on each side of the MEA. Silicone seal rings (25 mm OD, 16 mm ID, 1-mm thickness), are pressed on each side of the current collectors. Finally, a PTFE spacer ring (16 mm ID, 1-mm thickness) is placed as the last layer on the cathode side to provide adequate volume for compression. The glass cell, MEA holder, and O-rings are joined and held together using a screw clamping apparatus. It was designed to press the thick flat ends of the glass cell compartments together and ensured alignment of the glass compartments. This allows the application of a uniformly distributed force to seal the glass cell flanges to the MEA holder O-rings. All the experiments carried out at room temperature were conducted in ambient air (23°C ± 2°C). For experiments at 50°C and 75°C ± 1°C thermostated water was circulated within the double wall of the glass cell and the temperature controlled by an immersion circulator. Thermocouples were used to monitor the temperature in each of the glass cell compartments.

All electrochemical measurements were carried out at steady-state using a multistat (Solartron Analytical, Model 1470E) connected to a computer, using CoreWare software. A single junction mercury/mercurous sulfate reference electrode, placed in the anodic compartment, was used for all experiments at ambient temperature. At 50°C and 75°C, a double junction silver/silver chloride reference electrode (Ag/AgCl) was used, with 0.5 M H_2SO_4 in its outer compartment. All electrode potentials were quoted versus the standard hydrogen electrode. All electrolysis experiments were conducted in the galvanostatic, current-controlled mode. AC impedance measurements were carried out with an impedance/gain phase analyzer (Solartron Analytical, Model 1260 A) to obtain the MEA and solution resistance. All measured potential values were corrected for the ohmic drop before kinetic parameter information was extracted.

c. A third approach is to use an elementary PEM fuel cell, such as a direct methanol fuel cell (DMFC) hardware, working as a PEMEC [9].

A typical DMFC hardware, such as the one provided by ElectroChem, Inc. (ref.: EFC-05-02-DM), consists of a Pt/C negative electrode and a Pt–Ru (1/1 atomic ratio)/C positive electrode of 5 cm^2 surface area separated by a Nafion 117 membrane. The metallic loading is 2 mg cm^{-2} for the Pt/C cathode and 4 mg cm^{-2} for the Pt–Ru/C anode, respectively. The cell temperature can be controlled from 25°C to 85°C by circulating the reactant solutions with a peristaltic pump through a home-made thermostat and using the heating plates of the 5-cm^2 electrode surface area cell (Fig. 3.5):

By feeding the Pt/C electrode with pure hydrogen under 1 bar pressure at 25°C it is possible to measure the electrical characteristics versus the reversible hydrogen electrode (RHE) of the catalytic anode using the

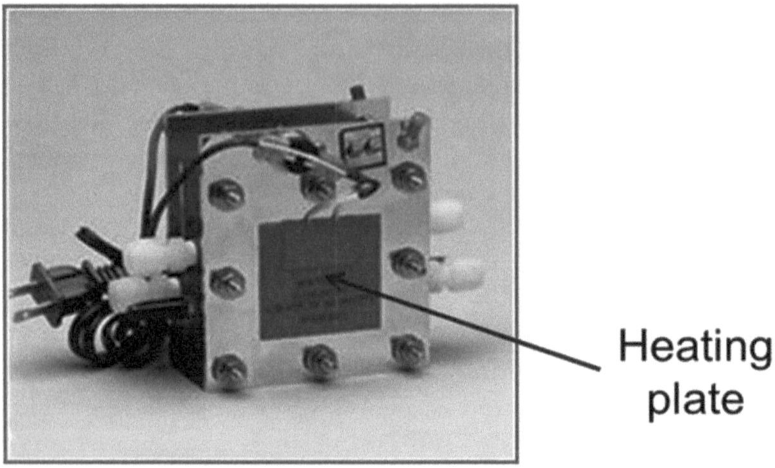

Figure 3.5 Photograph of the EFC-05-02-DM hardware provided by ElectroChem, Inc. *After [9].*

counter hydrogen electrode as a reference electrode. As an example, the Pt–Ru anode can be characterized by cyclic voltammetry at 25°C with a 0.5 M H_2SO_4 deaerated solution flowing through the anodic compartment with a liquid flow rate of 2 mL minute^{-1} at low potential sweep rates, between 1 and 10 mV second^{-1}. Then replacing the supporting electrolyte by the organic compound dissolved in 0.5 M H_2SO_4, the voltammetric curves characterizing its electrooxidation at the Pt–Ru anode can be recorded and analyzed with Tafel plots to determine the kinetics parameters at several temperatures.

For the electrochemical decomposition of the organic compound, the cell is polarized at a constant current density, j, using a potentiostat-galvanostat with the counter electrode linked to the reference electrode. The anode potential can be deduced from the cell voltage since the hydrogen evolution cathode can be used as a hydrogen reference electrode. In that way it was possible to obtain the electrical characteristics $E_a = f(j)$ of the oxidation of the organic compound at several concentrations and several working temperatures. These electrical characteristics, corrected from ohmic losses due to membrane and interface resistances, can then be analyzed leading to the cell resistance, the transfer coefficient, and the heat of activation.

The electrolysis experiments were carried out at a constant current intensity, I, by feeding the anode compartment with the organic solution

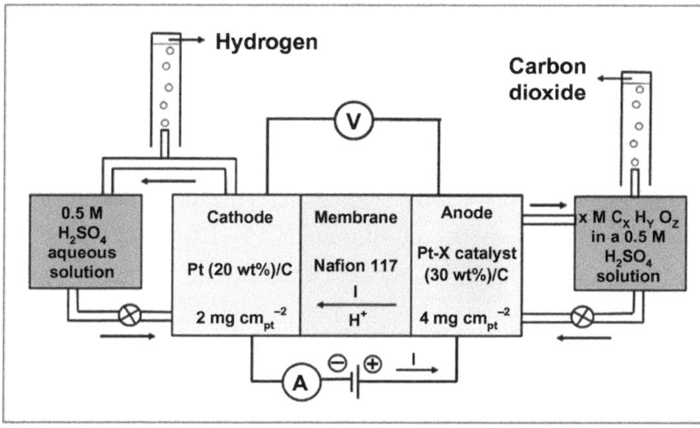

Figure 3.6 Experimental setup for the study of the electrochemical reforming of a hydrogen containing organic compound and for hydrogen evolution measurements. *After [9].*

at concentrations from 0.1 to 10 M and the cathode with an acidic solution (0.5 M H_2SO_4). The flow rate of the organic solution was chosen around 2 mL minute^{-1}; since it was observed that by increasing the flow rate to 40 mL minute^{-1} did not change the cell electrical characteristics, whereas that of the 0.5 M H_2SO_4 solution was fixed higher (37 mL minute^{-1}) to ensure that all the gaseous hydrogen produced is quickly evacuated in order to measure accurately the volume of evolved hydrogen. The volume of evolved hydrogen was measured by water displacement in a graduated glass burette connected to the cell. The experimental setup is schematized in Fig. 3.6.

3.4 Product analysis after electrolysis experiments

A high-performance liquid chromatograph (Dionex CHROMELEON) fitted with an isocratic pump, an autosampler, a UV detector, and a refractometer detector can be used to analyze quantitatively the reaction products at the outlet of the anodic side of the electrolysis cell. As an example, in the case of ethanol electroreforming, the volatile acetaldehyde (AAL) is analyzed after reaction with an excess of a 0.45 wt.% 2,4-dinitrophenylhydrazine in a 2.0 M HCl solution. The orange precipitate is then solubilized in ethyl acetate and analyzed by HPLC. The formed

CO_2 can react with a 1.0 M sodium hydroxide solution to form carbonate CO_3^{2-} anions. The CO_3^{2-} ions are detected by HPLC because of the formation of a negative peak. Two different columns can be used according to the products to be analyzed. For AAL, acetic acid (AA), and CO_3^{2-} analysis, a HPX-87H column was used with an eluent (0.01 M H_2SO_4 solution) flow of 0.6 mL minute^{-1}. For a hydrazone derivative, a C18-NH_2 column was used with an eluent (CH_3CN/H_2O 60:40) flow of 0.6 mL minute^{-1}. AAL and AA can be detected by the UV detector at a wavelength of 240 nm as well as by the refractometer detector, whereas CO_3^{2-} was detected using the refractometer only.

Semi-quantitative analyses of the electrooxidation products of glycerol and glucose were carried out after chrono-potentiometric measurements at different electrode potentials using a Varian Prostar HPLC equipped with a Transgenomc ICSep ICE-COREGEL 107 H column for the separation of organic acids, aldehydes, alcohols, and ketones and with an UV detector set at $\lambda = 210$ nm. The eluent was a 7 mM H_2SO_4 aqueous solution with a flow rate of 0.6 mL minute^{-1}. In the case of glycerol electrooxidation, DL-glyceraldehyde (purity $\geq$ 90%), dihydroxyacetone (97%), D-glyceric acid calcium salt dihydrate (99%), sodium β-hydroxypyruvate hydrate (purity $\geq$ 97%), tartronic acid (purity $\geq$ 97%), and sodium mesoxalate monohydrate (purity $\geq$ 98%) from Sigma–Aldrich were dissolved in 0.5 M aqueous NaOH, taking into account the acid-base reaction for compounds in their acid forms, and used to calibrate the chromatograms. For the analysis of the reaction products of glucose electrooxidation, reference solutions corresponding to the expected compounds were prepared from sodium gluconate, tartronic acid, 2-ketogluconic acid, 5-ketogluconic acid, glucuronic acid, oxalic acid, formic acid, tartric acid, and glycolic and glyoxylic acid from Sigma–Aldrich (purity $\geq$ 97%).

References

[1] M. Boutonnet, J. Kizling, P. Stenius, G. Maire, The preparation of monodisperse colloidal metal particles from microemulsions, Colloids Surf 5 (1982) 209–225. Available from: https://doi.org/10.1016/0166-6622(82)80079-6.

[2] C. Coutanceau, S. Brimaud, L. Dubau, C. Lamy, J.M. Léger, S. Rousseau, et al., Review of different methods for developing nanoelelectrocatalysts for the oxidation of organic compounds, Electrochim. Acta 53 (2008) 6865–6880. Available from: https://doi.org/10.1016/j.electacta.2007.12.043.

[3] L. Dubau, F. Hahn, C. Coutanceau, J.M. Léger, C. Lamy, On the structure effects of bimetallic PtRu electrocatalysts towards methanol oxidation, J. Electroanal. Chem. 554/555 (2003) 407–415. Available from: https://doi.org/10.1016/S0022-0728(03)00308-5.

[4] L. Dubau, C. Coutanceau, E. Garnier, J.M. Léger, C. Lamy, Electrooxidation of methanol at platinum-ruthenium catalysts prepared from colloidal precursors: atomic composition and temperature effect, J. Appl. Electrochem. 33 (2003) 419–429. Available from: https://doi.org/10.1023/A:1024491007321.

[5] W.S. Rasband, Image J, US National Institutes of Health, Bethesda, MD, USA. <http://rsbweb.nih.gov/ij/>, 2009 (accessed 25.10.18).

[6] H.K. Ju, S. Giddey, S.P.S. Badwal, R.J. Mulder, Electro-catalytic conversion of ethanol in solid electrolyte cells for distributed hydrogen generation, Electrochim. Acta 212 (2016) 744–757. Available from: https://doi.org/10.1016/j.electacta.2016.07.062.

[7] B. Beden, C. Lamy, Infrared reflectance spectroscopy, in: R.J. Gale (Ed.), Spectroelectrochemistry, Theory and Practice., Plenum Press, New York, 1988, ISBN: 978-1-4613-0985-7, pp. 189–261.

[8] T. Take, K. Tsurutani, M. Umeda, Hydrogen production by methanol–water solution electrolysis, J. Power Sources 164 (2007) 9–16. Available from: https://doi.org/10.1016/j.jpowsour.2006.10.011.

[9] C. Lamy, B. Guenot, M. Cretin, G. Pourcelly, Kinetics analysis of the electrocatalytic oxidation of methanol inside a DMFC working as a PEM electrolysis cell (PEMEC) to generate clean hydrogen, Electrochim. Acta 177 (2015) 352–358. Available from: https://doi.org/10.1016/j.electacta.2015.02.069.

[10] Z. Hu, M. Wu, Z. Wei, S. Song, P.K. Shen, Pt-WC/C as a cathode electrocatalyst for hydrogen production by methanol electrolysis, J. Power Sources 166 (2007) 458–461. Available from: https://doi.org/10.1016/j.jpowsour.2007.01.083.

[11] C.R. Cloutier, D.P. Wilkinson, Electrolytic production of hydrogen from aqueous acidic methanol solutions, Int. J. Hydrogen. Energy 35 (2010) 3967–3984. Available from: https://doi.org/10.1016/j.ijhydene.2010.02.005.

CHAPTER FOUR

Production of hydrogen by the electrocatalytic oxidation of low-weight compounds (HCOOH, MeOH, EtOH)

Contents

Instead of water several organic compounds, such as carboxylic acids, for example, formic acid, or low-weight alcohols, for example, methanol, ethanol, have been considered as convenient sources for the production of clean hydrogen by their electrocatalytic oxidation in a proton exchange membrane electrolysis cell (PEMEC).

First of all, the oxidation of formic acid, which is easily catalytically decomposed into hydrogen and carbon dioxide, is presented as a convincing proof of concept. Then low-weight alcohols are considered for their great industrial interest. For example, methanol is produced at a low cost from natural gas and can be completely decomposed into hydrogen and carbon dioxide, so that it is an excellent liquid storage of hydrogen. In contrast, ethanol is produced in great quantity and at a relatively low cost from biomass compounds, such as sugar cane, rootbeet, so that it can be an environmentally friendly source of hydrogen.

Production of Clean Hydrogen by Electrochemical Reforming of Oxygenated Organic Compounds.
DOI: https://doi.org/10.1016/B978-0-12-821500-5.00004-7

4.1 Electrocatalytic oxidation of formic acid

4.1.1 Introduction

The electrocatalytic oxidation of formic acid has been thoroughly investigated both from a fundamental aspect [1–6] and for its possible use in a direct formic acid fuel cell (DFAFC) [7,8]. However, very few investigations on the oxidation of formic acid to produce hydrogen by its electrochemical reforming can be found in the literature [9–11]. For the oxidation of formic acid in a DFAFC several oxidation, electrocatalysts have been shown to be very active, particularly Pd-based catalysts [8,11,12].

The oxidation reaction mechanisms are very well known involving two parallel pathways for the electrooxidation of formic acid into CO_2 [1–5], both leading to the exchange of two electrons per formic acid molecule. For example, with Pt-based catalysts, the "direct pathway" gives CO_2 without the formation of an adsorbed carbon monoxide intermediate:

$$\mathrm{Pt + HCOOH \rightarrow Pt + CO_2 + 2\,H^+ + 2\,e^-}$$

whereas the "indirect pathway" involves the formation of CO, which acts as a poisoning intermediate, but that can be further oxidized to CO_2 at higher potentials:

$$\mathrm{Pt + HCOOH \rightarrow Pt\text{-}CO + H_2O} \quad (4.1a)$$

$$\mathrm{Pt + H_2O \rightarrow Pt\text{-}OH + H^+ + e^-} \quad (4.1b)$$

$$\mathrm{Pt\text{-}CO + Pt\text{-}OH \rightarrow 2\,Pt + CO_2 + H^+ + e^-} \quad (4.1c)$$

where Pt represents a platinum catalytic site.

Both reaction paths correspond to the overall reaction:

$$\mathrm{Pt + HCOOH \rightarrow Pt + CO_2 + 2\,H^+ + 2\,e^-} \quad (4.2)$$

Similar mechanisms, particularly the direct pathway, were also observed on palladium-based catalysts [6,8], for example, Lu et al. showed that the addition of palladium to platinum led to enhance the direct reaction pathway [6]. Recent studies have suggested that palladium and palladium-based alloys have interesting activity compared to platinum for the electrooxidation reaction of formic acid [12–14].

4.1.2 Electrochemical decomposition of formic acid

For producing hydrogen by electrochemical reforming, formic acid is fed to the anodic compartment of a PEMEC, where its oxidation produces carbon dioxide and protons, that is:

$$HCOOH \rightarrow CO_2 + 2\,H^+ + 2\,e^- \tag{4.3}$$

Then the protons cross-over the protonic membrane and reach the cathodic compartment, where they are reduced to hydrogen according to reaction (2.2).

This corresponds to the electrochemical reforming of formic acid into hydrogen and carbon dioxide, according to the overall reaction:

$$HCOOH \rightarrow H_2 + CO_2 \tag{4.4}$$

The standard thermodynamic data associated with the reaction (4.3) or (4.4) can be calculated as follows:

$$\Delta H^{+0} = \Delta H^0_{CO_2} - \Delta H^0_{HCOOH} = -393.5 + 425.0 = 31.5\ \text{kJ}(\text{mol}_{HCOOH})^{-1}$$

$$\begin{aligned}\Delta G^{+0} &= \Delta G^0_{CO_2} - \Delta G^0_{HCOOH} = -394.4 + 361.4 \\ &= -33.0\ \text{kJ}(\text{mol}_{HCOOH})^{-1}\end{aligned}$$

This decomposition reaction needs external energy ($\Delta H > 0$), coming from the external electrical power source and of the surroundings, but in the case of formic acid, the decomposition is spontaneous, since $\Delta G < 0$. The corresponding theoretical cell voltage, under standard conditions, can be calculated from ΔG^{+0}, that is, $U^0_{cell} = (\Delta G^{+0} - \Delta G^{-o})/2F = E^+_a - E^-_c \approx E^+_a$ because $E^0_c \approx 0$ for the cathode in contact with evolving hydrogen under atmospheric pressure, so that $U^0_{cell} \approx \Delta G^{+0}/2F \approx -0.17$ V. Thus, assuming that some extra thermal energy is transferred from the surrounding (whose temperature can be controlled by a thermostat) to the electrolysis cell through the reversible heat transfer $\Delta Q_{rev} = T\,\Delta S$, the decomposition reaction of formic acid is a spontaneous process.

However, the relatively slow kinetics of the anodic reaction leads to high anodic overpotentials at high current densities (over 1 A cm^{-2}) necessary for high hydrogen production rates, that is, relatively high-cell voltages leading to high electrical energy consumption—see Eq. (2.11). In order to reduce these overpotentials to acceptable values, new electrocatalysts have been developed, such as Pt-based catalysts [15] or Pd-based catalysts, which have been recognized to be very active for the electrocatalytic oxidation of formic acid [8,13,14].

In this work [11], several Pd-based electrocatalysts dispersed on a carbon powder, such as Vulcan XC72, were prepared using the microemulsion method (Section 3.1.1).

4.1.3 Voltammetric results

In order to choose the best anode catalysts for the electrochemical oxidation of formic acid, cyclic voltammograms were recorded in a three-electrode cell with several Pd-based electrodes in a 0.5 M H_2SO_4 solution containing 0.01 M HCOOH (Fig. 4.1).

PdAu or PdPt alloys with a low content of Au or Pt (atomic ratios < 20%) display a good electrocatalytic behavior, in agreement with the results obtained at low overpotentials with a PtPd catalyst in a DFAFC as shown by Rice et al. [15] and with different PdAu catalysts in a classical electrochemical cell by Zhang et al. [16]. The voltammograms of the $Pd_{0.9}Au_{0.1}/C$ and $Pd_{0.8}Pt_{0.2}/C$ alloys display the forward and backward sweeps, which are remarkably quasi-superimposed (Fig. 4.1). This indicates that these catalytic surfaces are less sensitive to poisoning by adsorbed CO species resulting from the dissociative chemisorption of formic acid.

The electrocatalytic behavior of $Pd_{0.8}Pt_{0.2}/C$ is particularly interesting since the oxidation of formic acid begins at electrode potentials as low as 0.1 V versus reversible hydrogen electrode (RHE) leading to a peak

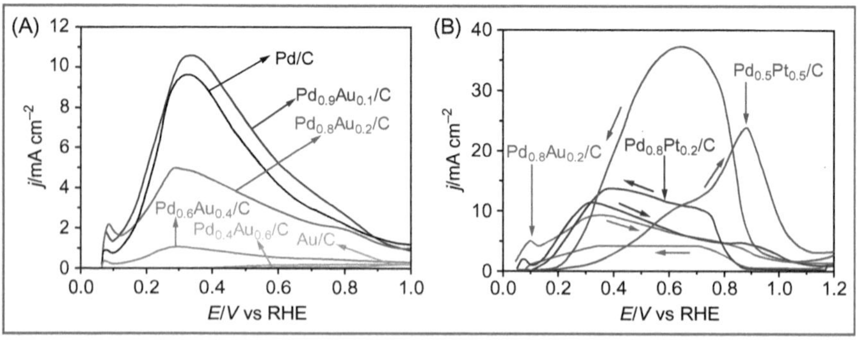

Figure 4.1 Voltammetric curves recorded during the oxidation of 10^{-2} M HCOOH in 0.5 M H_2SO_4 N_2-purged electrolyte on (A) different Pd_xAu_{1-x}/C catalysts and (B) comparison of the catalytic activity of Pd_xPt_{1-x}/C and $Pd_{0.8}Au_{0.2}/C$ catalysts ($T = 25°C$; $v = 50\ mV\ s^{-1}$).

current density of 12 mA cm^{-2} at 0.28 V. Similarly, the $Pd_{0.9}Au_{0.1}/C$ catalyst gives maximum current densities of about 11 mA cm^{-2} at 0.32 V versus RHE.

4.1.4 Electrolysis results

Therefore, long-term electrolyses of formic acid were undertaken in a PEMEC as described in Section 3.3(c). The electrolysis experiments were carried out at a constant controlled current intensity from 0.2 to 1 A, for example, a current density from 40 to 200 mA cm^{-2} with several Pd-based electrocatalysts ($Pd_{0.8}Au_{0.2}/C$, $Pd_{0.9}Au_{0.1}/C$, and $Pd_{0.8}Pt_{0.2}/C$). The cell voltage, U_{cell} at a fixed current intensity, was recorded as a function of time for 30 minutes of electrolysis.

In all experiments, the measured volume of hydrogen is a linear function of time (with correlation coefficients > 0.9986) showing clearly that the volume of evolved hydrogen does not depend on HCOOH concentration, nor on the nature of the electrode catalyst, but only on the current intensity I (Fig. 4.2), according to Eqs. (2.12) and (2.13).

Fig. 4.3 and Table 4.1 summarize all the results obtained with the three electrocatalysts ($Pd_{0.8}Au_{0.2}/C$, $Pd_{0.9}Au_{0.1}/C$, and $Pd_{0.8}Pt_{0.2}/C$) investigated here, where $dV/dt_{(average)}$ is the average value calculated with all the concentration (1−10 M) of formic acid. The experimental results are compared to the calculated values of dV_{H_2}/dt, the theoretical slope of the $V_{H_2}(30')$ during 30-minute electrolysis. The agreement is very good in all experiments, but with experimental values slightly higher than those calculated from the Faraday law—see Eqs. (2.12) and (2.13), which may come from the room temperature a little bit higher than 25°C. This positive deviation in the measured volume of evolved hydrogen could also be due to resistive heating in the cell during long-term electrolysis leading to an increase of the temperature of the exhaust gas.

For all experiments, the electrical energy needed to produce 1 Nm^3 of hydrogen was also evaluated, since it is only a function of the cell voltage U_{cell}, according to Eq. (2.11). As an example, the values obtained for U_{cell} after 30-minute electrolysis are given in Table 4.2 for the $Pd_{0.8}Au_{0.2}/C$ anode.

In all the results obtained, the amount of electrical energy is below 1.8 kWh $(Nm^3)^{-1}$ (since $U_{cell} < 0.8$ V), which is at least two to three times lower than the energy consumed for water electrolysis.

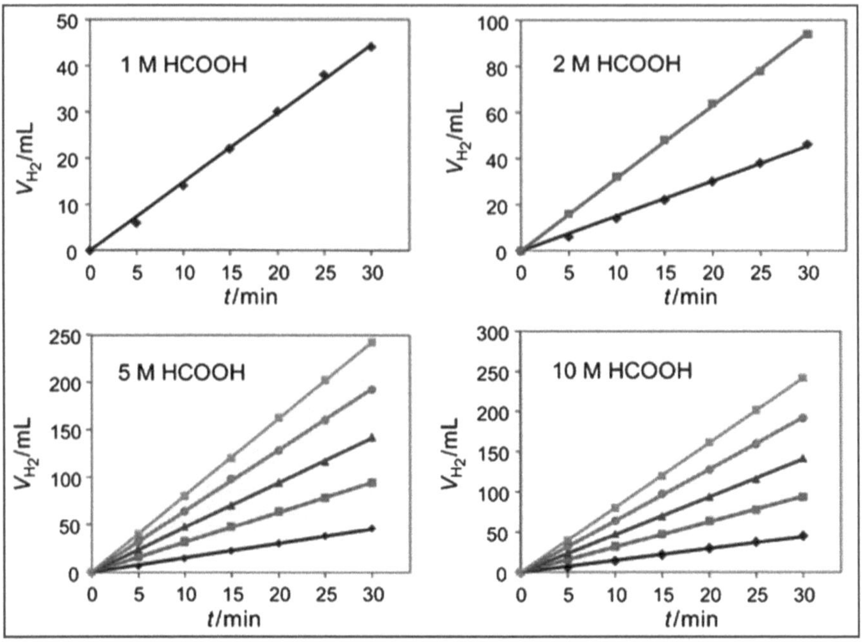

Figure 4.2 Hydrogen evolution at different current density for the PEMEC (0.5 M H_2SO_4, Pt/C, N117, $Pd_{0.8}Au_{0.2}/C$, × M HCOOH) at 25°C. (■) $j = 40\ mA\ cm^{-2}$, (●) $j = 80\ mA\ cm^{-2}$, (▲) $j = 120\ mA\ cm^{-2}$, (▼) $j = 160\ mA\ cm^{-2}$, and (◆) $j = 200\ mA\ cm^{-2}$.

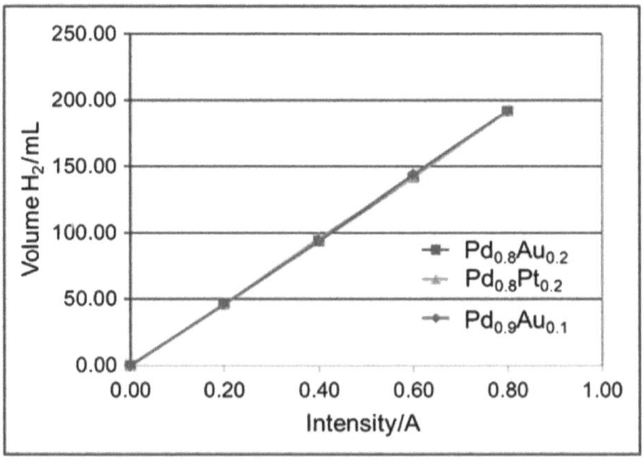

Figure 4.3 Hydrogen evolution as a function of the current intensity for the PEMEC with different anode catalyst (0.5 M H_2SO_4, Pt/C, N117, Pd_xM_{1-x}/C, × M HCOOH) at 25°C. (■) $Pd_{0.8}Au_{0.2}/C$, (▲) $Pd_{0.8}Pt_{0.2}/C$, (◆) $Pd_{0.9}Au_{0.1}/C$.

Table 4.1 Summary of the results obtained in a PEMEC for the different catalysts at 25°C with (HCOOH) = 5 M.

I/A	Theoretical values		$dV/dt_{(average)}$/cm^3 min^{-1}			$V_{H2}(30')_{experimental}$/$cm^3$		
	dV/dt/ $cm^3 min^{-1}$	$V_{H2(30')}$ / cm^3	$Pd_{0.9}Au_{0.1}$	$Pd_{0.8}Au_{0.2}$	$Pd_{0.8}Pt_{0.2}$	$Pd_{0.9}Au_{0.1}$	$Pd_{0.8}Au_{0.2}$	$Pd_{0.8}Pt_{0.2}$
0.2	1.52	46	1.51	1.51	1.51	46	46	46
0.4	3.04	92	3.17	3.15	3.2	94	94	96
0.6	4.56	137	4.78	4.70	4.70	144	142	142
0.8	6.09	183	6.40	6.41	6.41	192	192	192
1.0	7.61	228	8.05	8.07	Oscillation[a]	242	242	Oscillation[a]

[a]The oscillation of the cell voltage may result from the formation of adsorbed CO and its fast oxidation, as it was previously observed for the oxidation of formic acid at a rhodium electrode [17].

Table 4.2 Summary of the results for $Pd_{0.8}Au_{0.2}/C$ at 25°C with (HCOOH) = 5 M; U_{cell} is the PEMEC voltage and W_e is the electrical energy.

I/A	$U_{cell\ (t=0)}$/V	$W_{e_{(t=0)}}$/kWh $(Nm^3)^{-1}$	$U_{cell\ (t=30')}$/V	$W_{e_{(t=30')}}$/kWh $(Nm^3)^{-1}$
0.2	0.231	0.51	0.295	0.65
0.4	0.312	0.68	0.370	0.81
0.6	0.327	0.72	0.384	0.84
0.8	0.468	1.03	0.531	1.16
1.0	0.633	1.39	0.806	1.77

4.2 Electrocatalytic oxidation of low-weight alcohols (MeOH, EtOH)

The electrocatalytic oxidation of many alcohols, such as methanol, ethanol, ethylene glycol, glycerol, C3 mono-alcohols, and polyols, has been thoroughly studied in the literature since more than 50 years [18,19]. The main purposes were to establish the reaction mechanisms of their electrooxidation both in acid and alkaline media and to optimize the nature, composition, and structure of the best electrocatalysts to develop. In contrast, the electrocatalytic oxidation of many alcohols, for example, methanol, ethanol, ethylene glycol, was investigated in order to use them as a fuel in a direct oxidation fuel cell [20].

More recently, some alcohols and biomass compounds were considered as a hydrogen source for the production of hydrogen by their electrochemical decomposition [21–23]. One can find several reviews in the literature on the production of very pure hydrogen by the electrochemical reforming of alcohols and biomass compounds in an electrolysis cell similar to a PEM electrolyzer [21–24].

4.2.1 Electrochemical reforming of methanol

4.2.1.1 Introduction

Most of the previous studies on the electrochemical decomposition of methanol concerned the chemical process of hydrogen production by electrolysis [25–29] focusing on the effect of MeOH concentration (1–18 M) and temperature (30°C–90°C) on the rate of hydrogen evolution. None of these studies, except the work of Cloutier and Wilkinson [30], did investigate the kinetics of methanol oxidation under electrolysis cell conditions. Cloutier and Wilkinson used a two-compartment electrolytic glass cell in the static mode, separated by a membrane-electrode assembly (MEA) consisting in a Nafion 117 membrane on which are pressed the catalytic layers, Pt catalyst at the cathode for hydrogen evolution and Pt–Ru(1:1) catalyst at the anode for methanol oxidation. A mercurous sulfate electrode (MSE) was introduced in the anodic compartment as a reference electrode to control the anode potential by a potentiostat.

If most of these studies used a Nafion 117 membrane, a Pt–Ru anode and a Pt cathode to realize the MEA, they did not correlate clearly the hydrogen evolution rate with the current intensity flowing through the electrolysis cell as well as with the methanol concentration and cell temperature.

In a previous paper [31], we analyzed the reaction kinetics of methanol oxidation directly in the electrolysis cell using the hydrogen cathode as a reference electrode. In that way, it was possible to evaluate the overall resistance of the cell, that is, the electrolyte resistance + the charge transfer resistance of the hydrogen evolution reaction, the charge transfer coefficient ($\alpha \approx 0.5-0.6$) and the activation energy ($\Delta H^* \approx 50-60$ kJ mol^{-1}) for the electrochemical oxidation of methanol. This latter work was particularly innovative because of the use of a commercial direct methanol fuel cell (DMFC) hardware. Other previous fundamental studies were conducted in a basic laboratory two-compartment electrolyte glass cell in static mode. The electrodes were separated by a Nafion 117 membrane, on which were pressed the catalytic layers, and the anode potential was controlled by a potentiostat using an MSE in the anodic compartment as reference electrode [30].

In this study [31], we investigated the electrochemical decomposition of methanol for the production of very clean hydrogen. The electrolysis was performed in a DMFC hardware used as a PEMEC. The Pt/C hydrogen electrode of the DMFC served as a reference electrode, so that it was possible to analyze the kinetics of methanol oxidation at a catalytic Pt–Ru/C anode with the data extracted from the cell voltage versus current intensity corrected from ohmic losses. The electrolysis of methanol was carried out using a potentiostat in the galvanostatic mode with several methanol concentrations (0.1–10 M) at several temperatures (25°C–85°C).

4.2.1.2 Principle of methanol decomposition in a PEMEC

The principle of the electrochemical decomposition of methanol in a PEMEC involves its electrocatalytic oxidation similarly to what occurs at the anode of a DMFC [32]. In the anodic compartment, that is, the positive pole of the electrolysis cell, methanol is oxidized on the Pt–Ru catalyst, producing carbon dioxide and protons, according to reaction (4.5):

$$CH_3OH + H_2O \rightarrow CO_2 + 6\,H^+ + 6\,e^- \quad \text{anode reaction} \tag{4.5}$$

Then the protons, after crossing-over the protonic membrane, reach the cathodic compartment, that is, the negative pole of the electrolysis cell, where they are reduced to molecular hydrogen according to reaction (2.2)—see Section 2.2:

$$6\,H^+ + 6\,e^- \rightarrow 3\,H_2 \quad \text{cathode reaction} \quad (2.2)$$

This corresponds to the overall electrochemical decomposition of methanol into hydrogen and carbon dioxide, according to reaction (4.6):

$$CH_3OH + H_2O \rightarrow CO_2 + 3\,H_2 \tag{4.6}$$

This reaction is similar to methanol steam reforming, but it can occur at room temperature instead of elevated temperatures (200°C–400°C) for methanol reforming [33–35].

The thermodynamic data associated with reaction (4.5) can be calculated from the energy of formation of methanol, water, and carbon dioxide according to the following relations:

$$\Delta H^+ = \Delta H^f_{CO_2} - \Delta H^f_{MeOH} - \Delta H^f_{H_2O} \quad \text{and}$$
$$\Delta G^+ = \Delta G^f_{CO_2} - \Delta G^f_{MeOH} - \Delta G^f_{H_2O}$$

This gives $\Delta H^{+0} = +131.2$ kJ mol^{-1} and $\Delta G^{+0} = +9.3$ kJ mol^{-1} under standard conditions, corresponding to an external energy of 131.2/3 ≈ 43.8 kJ to be provided per mole of hydrogen produced, instead of 286 kJ mol^{-1} for water electrolysis and to an anode potential $E_a^+ = +\Delta G^{+0}/6\,F \approx 0.016$ V/SHE, that is, a cell voltage $U^0_{cell} = E_a^+ - E_c^- \approx 0.016$ V, since $E_c^- \approx 0$ V for the cathode potential of the hydrogen electrode versus the standard hydrogen electrode reference. For the DMFC, this corresponds to a standard cell voltage $E_{DMFC} = U^0_{H_2O} - U^0_{cell} = 1.229 - 0.016 = 1.213$ V.

This theoretical standard cell voltage ($U^0_{cell} \approx 0.016$ V) is very small compared to that of water electrolysis ($U^0_{cell} \approx 1.23$ V) so that the external energy, which is proportional to U^0_{cell} according to Eq. (2.11) in Section 2.2, theoretically needed to produce 1 mol of hydrogen by the electrochemical decomposition of methanol should be much smaller.

However, the relatively slow kinetics of the anodic reaction gives high anodic overpotentials, leading to greater cell voltages under high current densities over 1 A cm^{-2}, which are necessary for high hydrogen production rates, as described in Fig. 2.2 (see Chapter 2). In order to get a competitive cost of energy for the hydrogen production, the cell voltages have to be decreased down to acceptable values, through the development of new electrocatalysts with higher activity and selectivity. This involves decreasing the anodic overpotential through a detailed knowledge of the reaction mechanism of methanol oxidation and to find the best electrocatalysts in order to activate the rate-determining step (rds).

4.2.1.3 Reaction mechanisms of methanol oxidation

The mechanism of the electrooxidation of CH_3OH on Pt has been fully established, mainly after the identification of both reactive intermediates and adsorbed poisoning species [36]. In the first step, CH_3OH is dissociatively adsorbed at Pt-based catalysts by cleavage of C−H bonds leading to the so-called formyl-like species $-(CHO)_{ads}$. From this species, different steps can occur, but the dissociation of $-(CHO)_{ads}$ on Pt sites gives rapidly adsorbed CO, which is responsible for the electrode poisoning. This is the explanation of the rather poor performance of Pt catalysts due to the relatively high electrode potential necessary to oxidize CO_{ads}.

Some Pt/Ru-based tri-metallic electrocatalysts, such as Pt/Ru/Mo, give enhanced catalytic activity leading to a power density, in an elementary single DMFC, at least twice greater than that with a Pt/Ru catalyst [37]. This may be explained by the bifunctional theory of electrocatalysis, developed by Watanabe and Motoo [38], according to which Pt activates the dissociative chemisorption of CH_3OH to CO, whereas Ru activates and dissociates the water molecules, at lower potentials than Pt, leading to adsorbed hydroxyl species, OH_{ads}. A surface oxidation reaction between adsorbed CO and adsorbed OH becomes the rds. The reaction mechanism can be written as follows [39]:

$$Pt + CH_3OH \rightarrow Pt\text{-}CH_3OH_{ads} \quad (4.5a)$$

$$Pt - CH_3OH_{ads} \rightarrow Pt\text{-}CHO_{ads} + 3H^+ + 3\,e^- \quad (4.5b)$$

$$Pt - CHO_{ads} \rightarrow Pt - CO_{ads} + H^+ + e^- \quad (4.5c)$$

$$\text{either} \quad Pt + H_2O \rightarrow Pt - OH_{ads} + H^+ + e^- \quad (4.5d)$$

$$\text{or} \quad Ru + H_2O \rightarrow Ru - OH_{ads} + H^+ + e^- \quad (4.5e)$$

$$\text{either} \quad Pt - CO_{ads} + Pt - OH_{ads} \rightarrow 2\,Pt + CO_2 + H^+ + e^- \tag{4.5f}$$

$$\text{or} \quad Pt - CO_{ads} + Ru - OH_{ads} \rightarrow Pt + Ru + CO_2 + H^+ + e^- \tag{4.5g}$$

leading to the overall reaction $CH_3OH + H_2O \rightarrow CO_2 + 6\,H^+ + 6\,e^-$ (4.5)

An alternative path to the spontaneous formation of the poisoning species, according to reaction (4.5c), is the direct oxidation of CHO_{ads} species, with OH_{ads} species coming from the dissociation of H_2O (4.5d), according to the following reaction:

$$Pt - CHO_{ads} + Pt - OH_{ads} \rightarrow 2\,Pt + CO_2 + 2\,H^+ + 2\,e^- \tag{4.5h}$$

Parallel surface reactions, giving adsorbed formate, have also been observed:

$$Pt - (CHO)_{ads} + Pt - (OH)_{ads} \rightarrow Pt + Pt - (COOH)_{ads} + H^+ + e^- \tag{4.5i}$$

$$Pt - (CO_{ads}) + Pt - (OH)_{ads} \rightarrow Pt + Pt - (COOH)_{ads} \tag{4.5j}$$

then leading by further oxidation to the formation of CO_2:

$$Pt - (COOH)_{ads} \rightarrow Pt + CO_2 + H^+ + e^- \tag{4.5k}$$

This mechanism takes into account the formation of all the intermediate and reaction products, detected either by infrared reflectance spectroscopy [37] or liquid and gas chromatography [40]: formaldehyde through step (4.5b), formic acid through steps (4.5i) or (4.5j) and CO_2 through steps (4.5f), (4.5g), (4.5h), or (4.5k).

Thus, the crucial point is to determine the fate of the $-(CHO)_{ads}$ species. The different pathways for its oxidative removal are schematically summarized in Fig. 4.4. From this scheme, it appears that desorption and oxidation of the formyl species can follow different pathways through competitive reactions. This scheme summarizes the main problems and challenges to improve the kinetics of the electrocatalytic oxidation of CH_3OH.

On a pure Pt surface, step (4.5c) is spontaneously favored, and the formation of adsorbed CO is a fast process, even at low potentials. Thus, the coverage of adsorbed CO is high and explains the poisoning phenomena encountered at a Pt electrode. This poisoning species can be removed (by oxidation through step (4.5f) or (4.5g) into CO_2) only at potentials at

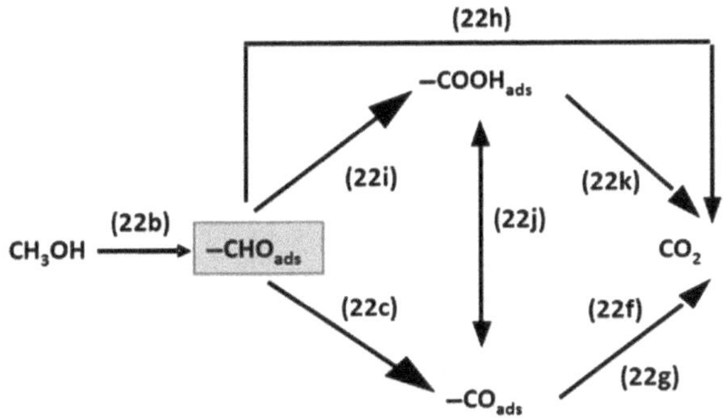

Figure 4.4 Schematic representation of the oxidation pathways of $-(CHO)_{ads}$ species.

which oxygenated species are formed at the electrode surface. At pure Pt, these oxygenated species arise from the dissociation of H_2O through step (4.5d), which occurs only at potentials more positive than 0.6 V [41]. Similarly the direct oxidation of $-(CHO)_{ads}$ into CO_2 through step (4.5h), or through step (4.5i) followed by step (4.5k) with the intermediate formation of $-(COOH)_{ads}$ species, needs again the presence of an extra oxygen atom, which can be provided only by the dissociation of H_2O at the catalytic surface.

Thus to lower the potential at which the dissociation of H_2O begins, a number of bimetallic and trimetallic Pt-based catalysts containing more easily oxidizable metals (Ru, Mo, Sn, Fe, Ni, etc.) have been investigated [41], among them carbon-supported Pt/Ru electrocatalysts lead to the best performance, so that they are currently used in DMFC stacks of a few watts to a few kilowatts. The atomic ratio between Pt and Ru, the particle size, and the metal loading of the carbon-supported anodes play a key role in their electrocatalytic behavior. Thus, commercial electrocatalysts, for example, from E-Tek, consist of (1:1) Pt/Ru catalysts dispersed on an electron-conducting substrate, for example, carbon powder such as Vulcan XC72 with a specific surface area of 200–250 $m^2\ g^{-1}$.

4.2.1.4 Experimental

All the electrolysis experiments were carried out in a DMFC hardware, provided by ElectroChem, Inc. (ref.: EFC-05-02-DM), consisting of a Pt/C negative electrode and a Pt–Ru (1/1 atomic ratio)/C positive electrode of 5 cm^2 surface area separated by a Nafion 117 membrane. The metal loading is 2 $mg\ cm^{-2}$ for the Pt/C cathode and 4 $mg\ cm^{-2}$ for the

Pt–Ru/C anode, respectively. The cell temperature was controlled from 25°C to 85°C by circulation of the reactant solutions with a peristaltic pump through a homemade thermostat and using the heating plates of the 5 cm^2 electrode surface area cell (see Fig. 3.5 in Section 3.3).

The electrolysis tests were performed by feeding the anodic compartment of the DMFC hardware with a 0.5 M H_2SO_4 solution containing methanol in concentrations ranging from 0.1 to 10 M and the cathode with a pure acidic solution (0.5 M H_2SO_4). The flow rate of the methanol solution was chosen around 2 mL $minute^{-1}$, since it was checked that increasing the flow rate up to 40 mL $minute^{-1}$ did not change the cell electrical characteristics, whereas that of the 0.5 M H_2SO_4 solution was fixed higher (35 mL $minute^{-1}$) to ensure that all the gaseous hydrogen produced is quickly evacuated in order to measure accurately its volume by water displacement in a graduated glass tube, for example, a burette connected to the cell (Fig. 4.5) [42].

The $U_{cell}(t)$ and $j(t)$ curves were recorded with a potentiostat–galvanostat used as a galvanostat in order to fix the current intensity applied to the cell from 5 to 800 mA, that is, a constant current density, j, from 1 to 160 mA cm^{-2}. The methanol anode potential was calculated from the cell voltage since the hydrogen evolution cathode can be used as a hydrogen reference electrode. In that way, it was possible to obtain the methanol oxidation electrical characteristics $E_{meth} = f(j)$ at several methanol concentrations from 0.1 to 10 M and several working temperatures from

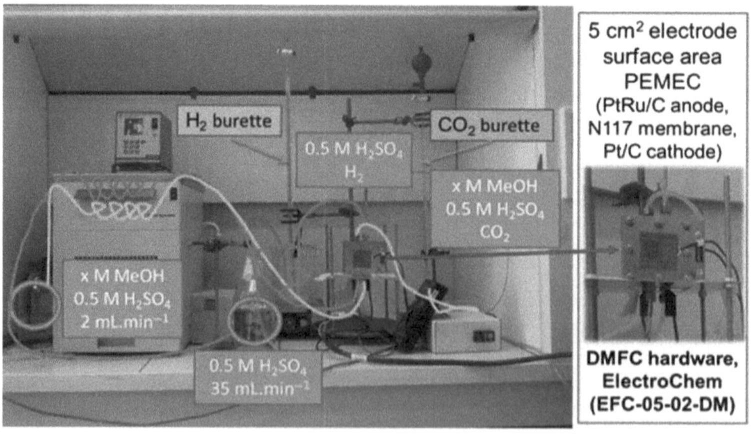

Figure 4.5 Experimental setup for the study of methanol electrolysis at the Pt–Ru (1:1)/C anode of the DMFC hardware and for hydrogen evolution measurements at the Pt/C cathode [42].

25°C to 85°C. These methanol oxidation characteristics, corrected from ohmic losses due to membrane and interface resistances, can be analyzed leading to the cell resistance, the charge transfer coefficient, and the heat of activation [31].

4.2.1.5 Voltammetric results

Cyclic voltammograms of the Pt–Ru electrode were recorded in a pure electrolyte solution (0.5 M H_2SO_4) at room temperature with sweep rates from 1 to 100 mV s^{-1} (Fig. 4.6).

The atomic hydrogen adsorption/desorption region on Pt, occurring between 0.05 and 0.40 V versus RHE, allowed to estimate the real surface area by calculating the quantity of electricity involved. The anodic surface area, associated to the oxidation of adsorbed hydrogen, corrected from the double layer capacity, corresponds to 187 mC, whereas that of the hydrogen adsorption corresponds to 193 mC, leading to an average value of 190 mC and an average surface area of about 900 cm^2, assuming 0.210 mC cm^{-2} of real Pt surface area.

The voltammograms, recorded at 1 mV s^{-1} and several temperatures (25°C–85°C) in the presence of 1 M methanol added to the supporting electrolyte (0.5 M H_2SO_4), are shown in Fig. 4.7 for a cathodic limit of 0.2 V versus RHE and an anodic limit of the sweep of 0.8 V in order to prevent Ru dissolution from the Pt–Ru anode. The general shape of the

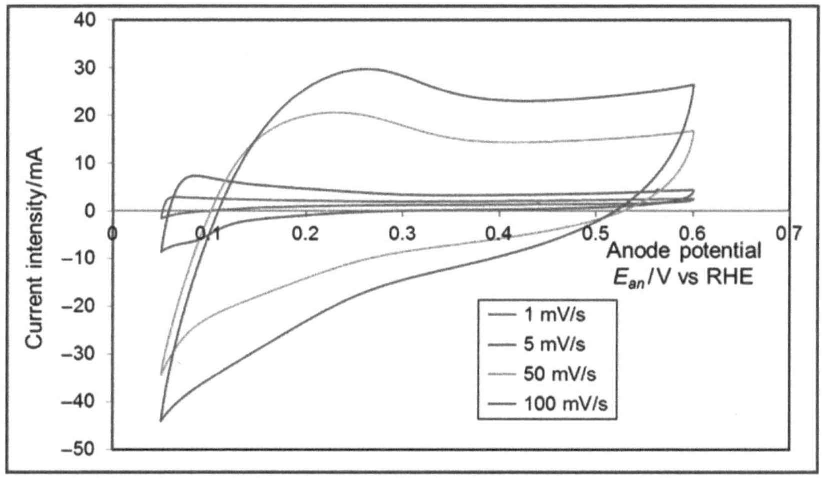

Figure 4.6 Cyclic voltammograms of the Pt–Ru/C anode in the supporting electrolyte (0.5 M H_2SO_4) at room temperature and at several sweep rates (1–100 mV s^{-1}).

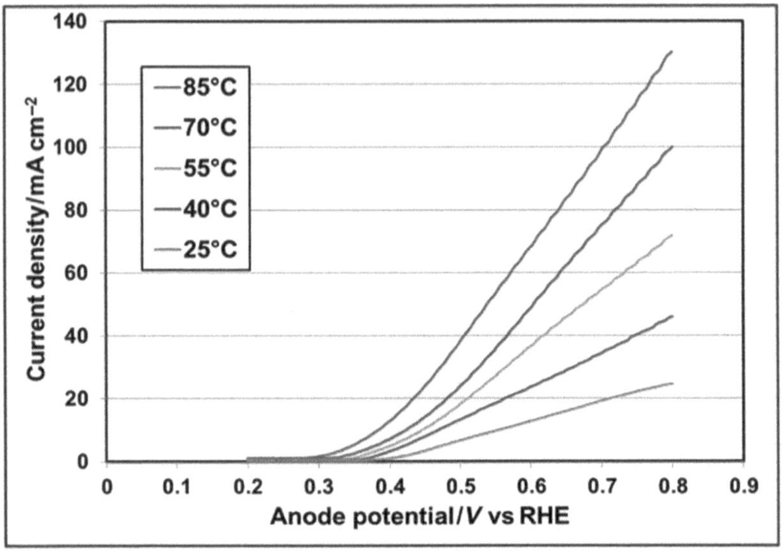

Figure 4.7 Polarization curves recorded at 1 mV s^{-1} and at several temperatures (from 25°C to 85°C) for the oxidation of 1 M methanol in 0.5 M H_2SO_4 at a PtRu(1:1)/C anode (4 mg cm^{-2}).

voltammograms is similar to that of voltammograms recorded in a three-electrode electrochemical cell for a Pt–Ru dispersed electrode [43].

4.2.1.6 Impedance Spectroscopy results

The electrochemical impedance spectroscopy measurements carried out at an anode potential of 500 mV in the presence of 2 M methanol in 0.5 M H_2SO_4 at several temperatures from 40°C to 85°C are given in Fig. 4.8.

From these curves, it is possible to evaluate the electrolyte and interface resistance R_e and the total resistance R_{total} including the charge transfer resistance of methanol oxidation. The high frequency resistance, R_e, the low frequency resistance, R_{total}, and the resistance evaluated from the polarization curves recorded under high-intensity current for hydrogen evolution, R_{exp}, are given for the temperature investigated (25°C–85°C) in Table 4.3 (see Section 4.2.1.7).

4.2.1.7 Electrolysis results

Electrolyses of several methanol solutions with concentrations ranging from 0.1 to 10 M were carried out in the DMFC hardware at different temperatures from 25°C to 85°C and at several constant controlled current densities from $j = 1$ to 100 mA cm^{-2}. In each case, the cell voltage

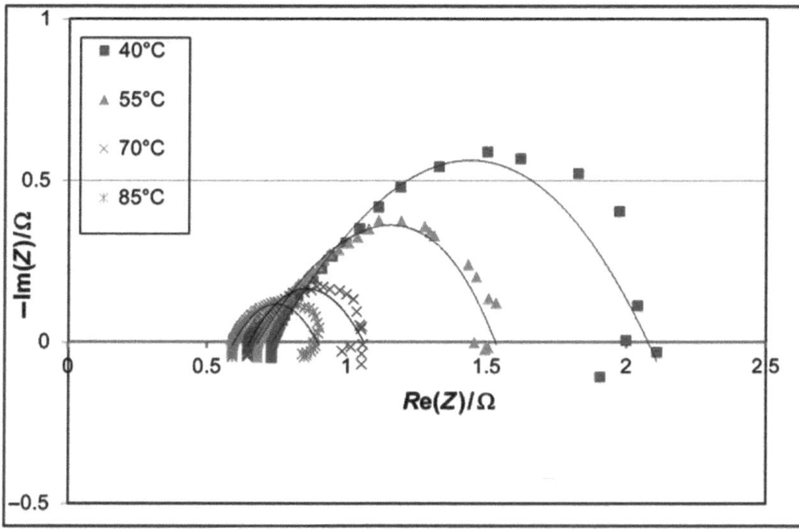

Figure 4.8 Impedance spectroscopy measurements of the DMFC hardware filled with a 0.5 M H_2SO_4 + 2 M CH_3OH solution at an anode potential of 500 mV and at different temperatures (T = 40°C–85°C, $f = 10^{-2}$ Hz to 10 kHz).

Table 4.3 Low-frequency resistance R_{total} and high-frequency resistance R_e as obtained from impedance spectroscopy and R_{exp} the resistance evaluated from the polarization curves, corrected from ohmic losses, that is, $U_{cell} - R_{exp} I$ versus ln(I), for 2 M CH_3OH at different temperatures.

T/°C	R_e/Ω	R_{total}/Ω	R_{exp}/Ω
25	1.86	4.9	1.9
40	0.74	2.1	1.36
55	0.68	1.5	1.14
70	0.65	1.05	0.95
85	0.60	0.88	0.75

U_{cell} and the volume of evolved hydrogen were recorded as a function of electrolysis time (t = 0–20 minutes).

The cell voltage, U_{cell}, reaches rapidly a steady-state value, usually after 15–20 minutes of electrolysis (Fig. 4.9), so that in the whole following experiments, the data corresponding to steady-state polarization curves were taken after 20 minutes of electrolysis. The cell voltage is lower at lower current intensity and higher temperatures showing the activation of methanol oxidation rate by increasing the temperature.

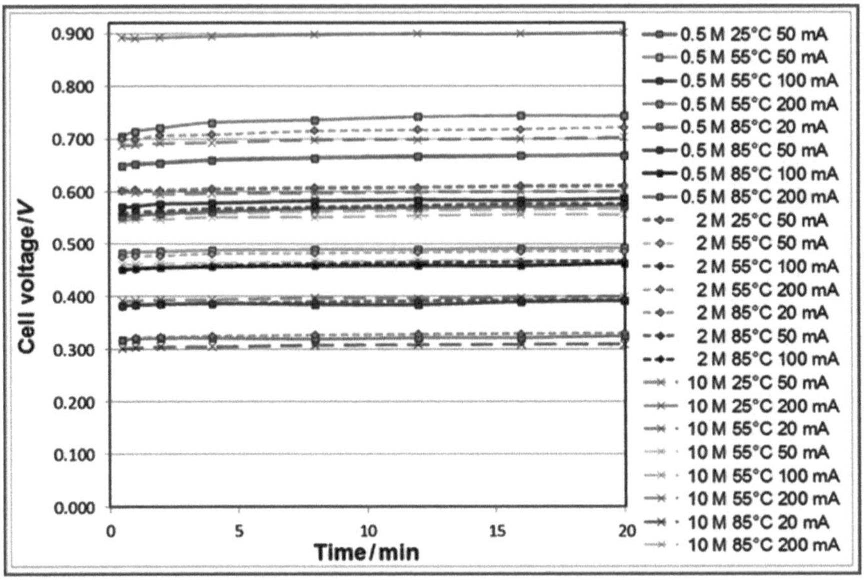

Figure 4.9 Examples of cell voltage versus time plots for several methanol concentrations (0.5–10 M), working temperatures (25°C–85°C) and current intensities (20–200 mA). The steady state is reached after 20 minutes of electrolysis.

The polarization curves as a function of current density, $U_{cell} = f(j)$, taken when the steady state was reached, usually after 20 minutes of electrolysis at a given current intensity I, are summarized in Fig. 4.10 for the different methanol concentrations and working temperatures.

These curves are relatively independent of methanol concentration, except for 0.1 M, which is not given in Fig. 4.10, at a given temperature, but the polarization curves depend greatly on the cell temperature for a given methanol concentration. At a given cell voltage, for example, $U_{cell} = 600$ mV, for a 2 M methanol solution, the current density increases from 10 mA cm^{-2} at 25°C to 43.5 mA cm^{-2} at 85°C. This is the consequence of temperature activation of methanol oxidation with an activation energy $\Delta H^* \approx 55$ kJ mol^{-1} [31]. These polarization curves are similar to those encountered in the electrocatalytic oxidation of methanol on a Pt–Ru electrode in a DMFC [44].

The experimental resistance, R_{exp}, which is unknown, has been determined by plotting $U_{cell} - R_{exp}\, I$, calculated for each value of I with an arbitrary value of R_{exp}, as a function of $\ln(I)$ assuming different reasonable values of R_{exp}, of the order of a few ohms, until a linear plot was obtained

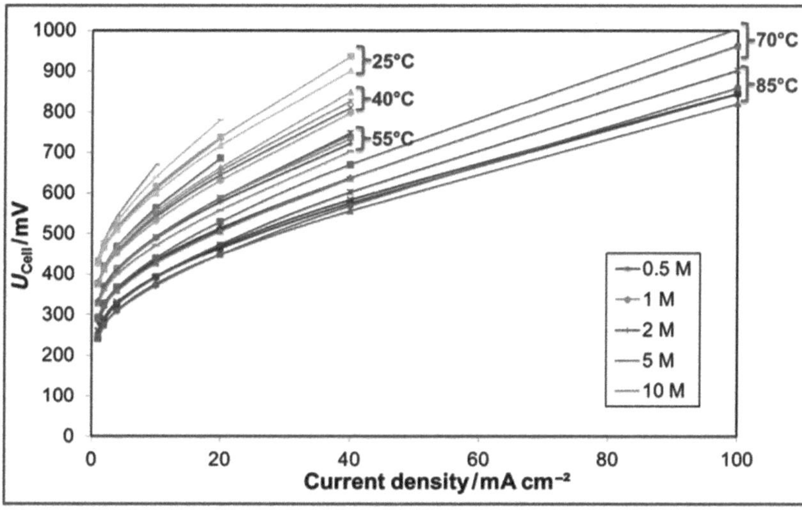

Figure 4.10 Polarization curves of methanol oxidation for several methanol concentrations (0.5–10 M) and at different working temperatures (25°C–85°C).

with a good fitting coefficient ($R^2 > 0.999$)—see Eq. (2.10) in Section 2.2.

As an example, the results obtained for a 2 M methanol solution are shown in Fig. 4.11 for an investigated temperature range from 25°C to 85°C. The different values of the total equivalent resistance R_{exp}, leading to a good linear relationship between $U_{cell} - R_{exp}\,I$ and $\ln(I)$, are given in Table 4.3, together with the results of impedance measurements.

These results are quite similar to those obtained from electrochemical impedance spectroscopy leading to R_{exp} values between 1.9 and 0.75 Ω for T varying from 25°C to 85°C (see Section 4.2.1.6).

4.2.1.8 Measurement of the hydrogen evolution rate

The hydrogen evolution rate was determined by measuring the volume of evolved hydrogen as a function of time at several current intensities ($I = 1-500$ mA) during the electrolysis of different concentration of methanol (0.1–10 M).

In all experiments, the measured volume of hydrogen is a linear function of time—see Fig. 4.12(A) with correlation coefficients $R^2 >$ 0.9968—showing clearly that the volume of evolved hydrogen does not depend on the methanol concentration, neither on the cell temperature nor on the nature of the anode catalyst, but only depends on the quantity

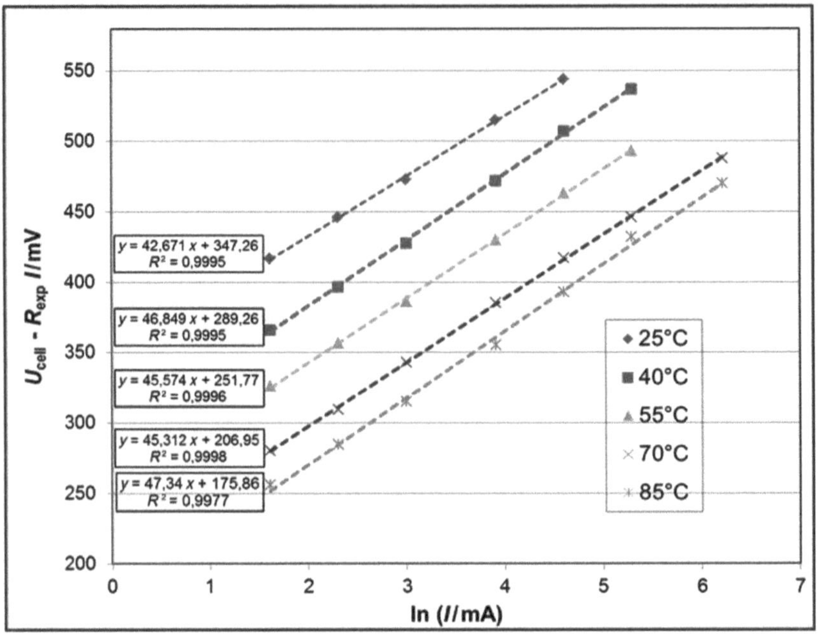

Figure 4.11 Plots versus the common logarithm of the current intensity, *I*, of the cell voltage corrected from an optimized ohmic drop (determined to obtain a linear correlation between $U_{cell} - R_{exp}\ I$ and ln *I*) for a 2 M CH_3OH solution at different temperatures (from 25°C to 85°C).

of electricity involved, $\Delta q = I\ \Delta t$, that is, on the electrolysis time Δt and the current intensity I (Fig. 4.12(B)), as a linear function according to Eq. (2.13)—see Section 2.2.

The three plots in Fig. 4.12(A), obtained at 50 mA are quite superimposed, in accordance with the Faraday's law—see Eq. (2.13) in Section 2.2.

Tables 4.4 and 4.5 give the experimental volume of evolved hydrogen, the measured cell voltage U_{cell}, the experimental cell resistance R_{exp}, the cell voltage corrected from ohmic losses, and the electrical energy W_e needed to produce 1 Nm^3 of hydrogen, according to Eq. (2.11) in Section 2.2, as a function of methanol concentration (Table 4.4) and cell temperature (Table 4.5). The cell voltage, corrected from ohmic losses (i.e., $U_{cell} - R_{exp}\ I$), was calculated using the experimental cell resistance, R_{exp}, estimated from the kinetics analysis of methanol electrooxidation in the DMFC hardware [31].

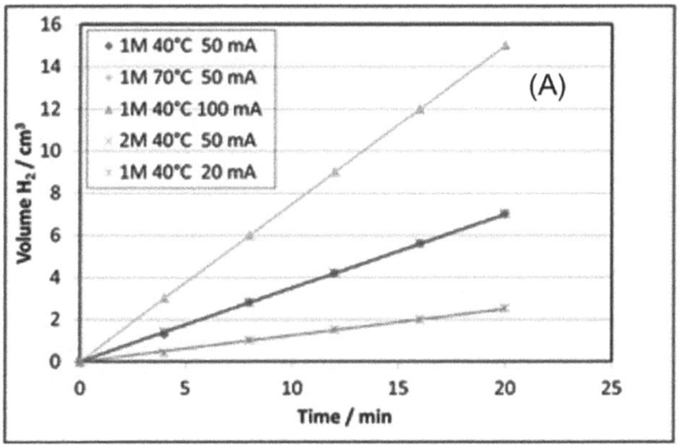

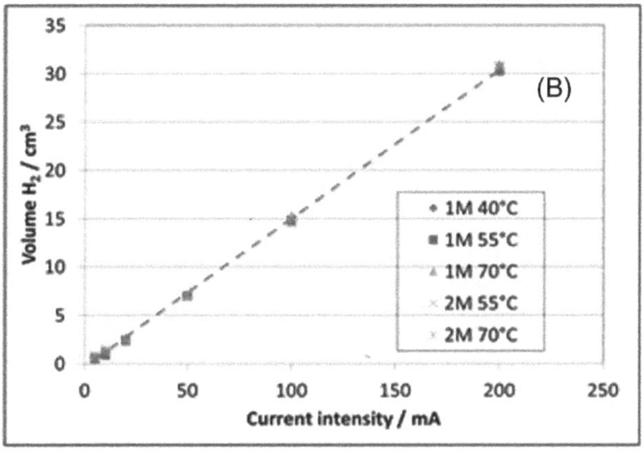

Figure 4.12 Examples of curves giving the volume of evolved hydrogen for two methanol solutions (1 and 2 M) as a function of (A) the electrolysis time Δt at two working temperatures (40°C and 70°C) and different current intensities(20, 50, and 100 mA) and (B) the current intensity *I* after 20 minutes of electrolysis at three working temperatures (40°C, 55°C, and 70°C).

As observed in Fig. 4.12(B) and in Tables 4.4 and 4.5, the volume of hydrogen evolved after 20 minutes of electrolysis only depends on the current intensity irrespective of methanol concentration, working temperature, cell voltage, and nature of the anode catalyst. However, the electrical energy consumed depends greatly on the working temperature and on the nature of the anode catalyst, since it is related to the cell voltage, that is, to the kinetics of the anodic process at a given current intensity.

In all the results obtained, the amount of electrical energy used to produce clean hydrogen by the electrochemical decomposition of methanol

Table 4.4 Experimental volume of generated H_2 compared to the theoretical volume of H_2 after 20 min electrolysis at 25°C and 1 atm.

(MeOH)/mol L^{-1}	Volume of evolved H_2/cm^3	Cell voltage/mV	R_{exp}/Ω	Corrected cell voltage/mV	W_e/kWh $(Nm^3)^{-1}$
0.1	15.3	651	2.2	431	0.94
0.5	14.6	509	0.89	420	0.92
1	15.3	512	0.93	419	0.92
2	14.5	512	0.95	417	0.91
5	14.6	528	1	428	0.94
10	14.7	505	0.98	407	0.89

(V_{H_2theo} = 15.21 cm^3), cell voltage, interfacial resistance R_{exp}, cell voltage corrected from ohmic losses, and electrical energy W_e, as a function of methanol concentration for 20 min electrolysis at 70°C and I = 100 mA with a PtRu(1:1)/C anode.

Table 4.5 Experimental volume of generated H_2 (V_{H_2theo} = 15.21 cm^3 at 25°C and 1 atm. after 20 min electrolysis), cell voltage, interfacial resistance R_{exp}, cell voltage corrected from ohmic losses, and electrical energy W_e, as a function of temperature for the electrolysis of 2 M methanol at 100 mA in a 5 cm^2 PEMEC with a PtRu(1:1)/C anode.

T/°C	Volume of evolved H_2/cm^3	Cell voltage/mV	R_{exp}/Ω	Corrected cell voltage/mV	W_e/kWh $(Nm^3)^{-1}$
25	14.8	734	1.9	544	1.19
40	14.6	643	1.36	507	1.11
55	14.9	577	1.14	463	1.01
70	14.5	512	0.95	417	0.91
85	15.0	468	0.75	393	0.87

is below 1.2 kWh $(Nm^3)^{-1}$, since corrected $U_{cell} < 0.55$ V, which is at least three to four times lower than the energy consumed for water electrolysis.

The energy efficiency of the process ε can be calculated [45] as the ratio between the minimum energy needed to decompose completely a methanol molecule leading to 3 mol of hydrogen, that is, $\Delta H^{+0} = 131.2\ 10^3$ J mol^{-1} under standard conditions, and the energy effectively used, that is, $\Delta H^{+} = \Delta H^{+0} + 6F\ \eta_{loss}$ with $\eta_{loss} \approx U_{cell} - U^0_{cell} \approx U_{cell}$ since $U^0_{cell} = 0.016\ V \approx 0$. Taking the value of $U_{cell} = 0.544$ V, corrected from ohmic losses, at 25°C and 100 mA for 2 M methanol (Table 4.5), this gives the following efficiency:

$$\varepsilon_{25} = \Delta H^{+0}/(\Delta H^{+0} + 6F\ U_{cell}) = 131.210^3/(131.2\ 10^3 + 6 \times 96485 \times 0.544) = 0.294 \approx 30\%$$

At 70°C for a 2 M methanol solution, we have $\Delta H^{+0} = 129.5\ 10^3$ J mol^{-1} (see Table 2.2) and a corrected $U_{cell} = 0.417$ V at 100 mA (Table 4.5), so that:

$$\varepsilon_{70} = 129.510^3/(129.5\ 10^3 + 6 \times 96485 \times 0.417) = 0.349 \approx 35\%$$

These values of energy efficiency are about half of those usually encountered in water electrolysis process, that is, 60%–70% at 70°C–80°C, but the electrical energy consumed is much smaller, that is, 1–1.2 kWh $(Nm^3)^{-1}$ compared to 5–6 kWh $(Nm^3)^{-1}$ for water electrolysis.

4.2.2 Electrochemical reforming of ethanol

4.2.2.1 Introduction

The electrochemical oxidation of ethanol has been extensively studied since many years, particularly for its use in a direct ethanol fuel cell (DEFC), since it is a liquid fuel easily transported and stored, with a rather high energy density of 8 kWh kg^{-1}. Furthermore, it can be produced from biomass feedstock, that is, dedicated crops, grain, corn, rootbeet, or from different wastes by enzymatic hydrolysis and fermentation [20,46,47].

More recently, ethanol and bio-ethanol have been considered as very convenient sources of clean hydrogen for low-temperature fuel cells, particularly when produced by electrochemical reforming in a PEMEC [24,47–49]. Caravaca et al. studied the electrochemical reforming of ethanol–water solutions in a PEMEC fitted with a bimetallic alloy Pt–Ru anode with a metal loading of 3 mg PtRu cm^{-2}, a Pt cathode with a loading of 0.5 mg Pt cm^{-2}, and a Sterion membrane (David Fuel Cell Components SL, Madrid, Spain) of 185-μm thickness as the electrolyte. They obtained a maximum current density of 0.25 A cm^{-2} at 1.1 V and 80°C, corresponding to an electrical energy consumption of c.2.6 kWh $(Nm^3\ H_2)^{-1}$ [48]. Similar results were obtained with a bio-ethanol solution from the wine industry [49].

The most important point is to develop very active and specific electrocatalysts working under the conditions of a PEMEC, which experiences a very high acidity. Thus, the main challenge is to conceive specific catalysts, which are mainly based on platinum group metals, particularly Pt-based electrocatalysts dispersed on high surface area, for example, Vulcan XC72, similar to that used in the DEFC [50].

The feasibility of the production with higher rate of clean hydrogen by the electrolysis of ethanol in a PEMEC has been investigated. Different Pt-based catalysts have been considered for the anodic oxidation of ethanol leading to high reaction rates at relatively low overvoltages, of the order of 0.8–0.9 V at 100 mA cm^{-2}, as experienced in a DEFC [51].

4.2.2.2 Reaction mechanisms of ethanol oxidation

To develop efficient electrocatalysts allowing decreasing the cell voltage of the electrolysis cell, it is important to recall the detailed reaction mechanisms of ethanol oxidation in order to know the rds, the kinetics of which has to be improved by a right choice of the catalyst nature and structure.

The electrochemical oxidation of C_2H_5OH has been extensively studied at Pt electrodes [37,39,50–57]. The first step is the dissociative adsorption of C_2H_5OH, either via an O-adsorption or a C-adsorption process [53,54], to form acetaldehyde (AAL) at potentials lower than 0.6 V versus RHE [52]. At higher potentials, the H_2O dissociation on Pt produces OH_{ads} necessary to provide an extra oxygen atom to realize the complete oxidation of C_2H_5OH to CO_2 with the cleavage of the −C−C− bond.

Pure Pt smooth electrodes are rapidly poisoned by some strongly adsorbed intermediates, such as CO, resulting from the dissociative chemisorption of the molecule, as shown by the first experiments in *infrared reflectance spectroscopy* [58]. Both kinds of adsorbed CO, either linearly bonded or bridge-bonded to the Pt surface, are observed. Moreover, other adsorbed species have been identified by *IR reflectance spectroscopy* [58], and reaction intermediates and by-products, such as AAL and acetic acid (AA), by chromatographic analysis [53] and by *differential electrochemical mass spectrometry* (DEMS) [53,54,59]. Voltammetric results, completed by the different spectroscopic and chromatographic results, allowed proposing a detailed reaction mechanism of ethanol oxidation, involving parallel and consecutive reactions, on Pt-based electrodes, such as Pt−Sn catalysts, where the key role of the adsorption steps was underlined.

Since it is relatively difficult to oxidize completely ethanol into carbon dioxide at room temperature, two main parallel reaction paths, leading to the formation of intermediate products, for example, AAL or AA, have been considered:

$$CH_3 - CH_2OH \rightarrow CH_3 - CHO + 2\,H^+ + 2\,e^- \quad \text{(reaction rate } \nu_1\text{)} \tag{4.7}$$

or

$$CH_3 - CH_2OH + H_2O \rightarrow CH_3 - COOH + 4\,H^+ + 4\,e^- \text{(reaction rate } \nu_2\text{)} \tag{4.8}$$

Depending on the kind of electrocatalyst used and on the electrode potential at which the reaction does occur, the reaction rate ν_1 or ν_2 controls the rds leading predominantly either to AAL or to AA.

Further oxidation to CO_2 needs the presence of extra oxygen atoms provided by the oxidation of adsorbed water, according to the bifunctional mechanism [60]. This process is favored on Pt-based electrocatalysts, such as Pt-X with X = Sn, Ru, and so on, at potentials from 0.4 to 0.6 V versus RHE, lower than those with Pt alone, for example, 0.6–0.8 V versus RHE.

Voltammetric results, completed by the different spectroscopic and chromatographic analyses, allowed proposing a detailed reaction mechanism of C_2H_5OH oxidation, involving parallel and consecutive oxidation reactions, on Pt-based electrodes, for example, such as Pt–Sn catalysts, where the key role of the adsorption steps was underlined [46].

$$Pt + CH_3 - CH_2OH \rightarrow Pt - (CHOH-CH_3)_{ads} + H^+ + e^- \tag{4.9a}$$

$$Pt - (CHOH-CH_3)_{ads} \rightarrow Pt + CHO - CH_3 + H^+ + e^- \quad \text{at } E < 0.6\text{ V vs RHE} \tag{4.9b}$$

As soon as AAL is formed, it can adsorb on Pt sites leading to a $-(CO-CH_3)_{ads}$ species (adsorbed acetyl):

$$Pt + CHO - CH_3 \rightarrow Pt - (CO-CH_3)_{ads} + H^+ + e^- \tag{4.9c}$$

At potentials higher than 0.6 V versus RHE, the dissociative adsorption of H_2O occurs on Pt providing $-OH$ adsorbed species, that is, $Pt + H_2O \rightarrow Pt-(OH)_{ads} + H^+ + e^-$, able to oxidize further the adsorption residues of C_2H_5OH, leading to AA, as follows:

$$\text{either} \quad Pt - (CHO - CH_3)_{ads} + Pt - (OH)_{ads} \rightarrow 2\,Pt + CH_3 - COOH + H^+ + e^- \tag{4.9d}$$

$$\text{or} \quad Pt-(CO-CH_3)_{ads} + Pt-(OH)_{ads} \rightarrow 2\,Pt + CH_3-COOH \tag{4.9e}$$

The presence of Sn in the catalytic layer can activate the dissociation of H_2O, thus providing $-(OH)_{ads}$ species at lower potentials than on Pt, according to reaction (4.9f):

$$Sn + H_2O \rightarrow Sn-(OH)_{ads} + H^+ + e^- \tag{4.9f}$$

so that adsorbed AAL species can react with adsorbed OH species to give AA at lower potentials, according to:

$$Pt-(CO-CH_3)_{ads} + Sn-(OH)_{ads} \rightarrow Pt + Sn + CH_3-COOH \quad \text{at } E < 0.6\ \text{V vs RHE} \tag{4.9g}$$

Further oxidation to CO_2 is usually difficult on pure Pt electrodes at room temperature, since it is difficult to break the $-C-C-$ bond. However, CO acting as a poisoning species and CO_2 were clearly observed by *IR spectroscopy* [58] (and CO_2 was detected by DEMS [53,54]) and by gas chromatography, whereas some traces of CH_4 were observed at low potential, that is, $E < 0.4$ V versus RHE, by DEMS [53,54]. This may be explained by the following mechanism involving the dissociation of $-(CO-CH_3)_{ads}$ by breaking the C–C bond, as observed by *IR reflectance spectroscopy* [58]:

$$Pt + Pt-(CO-CH_3)_{ads} \rightarrow Pt-(CO)_{ads} + Pt-(CH_3)_{ads} \quad \text{at } E > 0.3\ \text{V vs RHE} \tag{4.9h}$$

$$2\,Pt + H_2O \rightarrow Pt-H_{ads} + Pt-OH_{ads} \tag{4.9i}$$

$$Pt-(CH_3)_{ads} + Pt-H_{ads} \rightarrow CH_4 + 2\,Pt \quad \text{at } E < 0.4\ \text{V vs RHE} \tag{4.9j}$$

Then, oxidation of adsorbed CO species occurs, as was shown by *FTIR reflectance spectroscopy* and CO stripping experiments [51]:

$$\text{either} \quad Pt-(CO)_{ads} + Pt-(OH)_{ads} \rightarrow 2\,Pt + CO_2 + H^+ + e^- \quad \text{at } E > 0.6\ \text{V vs RHE}(4.5f)$$

$$\text{or} \quad Pt-(CO)_{ads} + Sn-OH_{ads} \rightarrow Pt + Sn + CO_2 + H^+ + e^- \quad \text{at } E < 0.6\ \text{V vs RHE} \tag{4.9k}$$

The reaction mechanism of ethanol oxidation on Pt-based electrodes can be summarized by the scheme in Fig. 4.13 [61,62]:

In this mechanism, the adsorbed acetyl plays a key role, and its further oxidation is favored by the addition to platinum of metal atoms, such as Ru, Sn, Mo, and so on, more oxophilic than Pt at lower potentials.

This mechanism can explain the higher efficiency of PtSn in forming AA compared with Pt at low potentials, for example, $E < 0.5$ V versus RHE, as was shown by electrolysis experiments [63]. Moreover, adsorbed OH species formed at lower potentials on Sn atoms can oxidize $-(CO-CH_3)_{ads}$ adsorbed species to CH_3-COOH or CO species to CO_2, according to the bifunctional mechanism [60].

Many Pt-based electrocatalysts, including Pt/X alloys with X = Re, Rh, Ru, Sn, and so on, and dispersed nanoparticles, have been investigated to enhance the catalytic activity of ethanol electrooxidation (Fig. 4.14) [51].

This figure shows clearly that Pt–Sn electrocatalysts are much more active at relatively lower potentials, that is, from 150 mV versus RHE, than pure Pt for which the oxidation of ethanol starts at 300 mV versus RHE). Indeed, it has already been well established that the modification of Pt by Sn led to a great improvement in the catalytic activity toward the ethanol electrooxidation reaction, in particular for Sn/Pt ratios lower than 0.3 [51,64]. Rousseau et al. also showed that addition of ruthenium to a Pt–Sn anode catalyst allowed achieving a maximum power density in a DEFC twice higher than that obtained with a Pt–Sn anode catalyst alone, giving c.30 mW cm^{-2} with a $Pt_{0.9}Sn_{0.1}/C$ anode to c.60 mW cm^{-2} with a $Pt_{0.86}Sn_{0.1}Ru_{0.04}/C$ anode [50]. The modification of PtSn by Rh has also been investigated because Rh increases CO_2 yield by facilitating the C–C bond cleavage [65–67]. However, the addition of Rh to PtSn does not increase the current of ethanol electrooxidation [67,68], conversely to the addition of ruthenium.

The composition of the catalysts considered here [63] was chosen according to previous studies on ethanol electrooxidation [50,51]. Several Pt-based electrocatalysts, such as Pt/C, $Pt_{0.9}Sn_{0.1}/C$, and $Pt_{0.86}Sn_{0.1}Ru_{0.04}/C$, were synthesized using the Bönnemann method (see Section 3.1.2) since it was clearly shown that these catalysts led to higher activity toward the electrooxidation of ethanol. Cyclic voltammograms (Fig. 4.15) recorded in a three-electrode cell containing a 0.1 M $HClO_4$ + 0.1 M C_2H_5OH solution with these three anodes, for example, Pt/C, $Pt_{0.9}Sn_{0.1}/C$, and $Pt_{0.86}Sn_{0.1}Ru_{0.04}/C$, confirmed

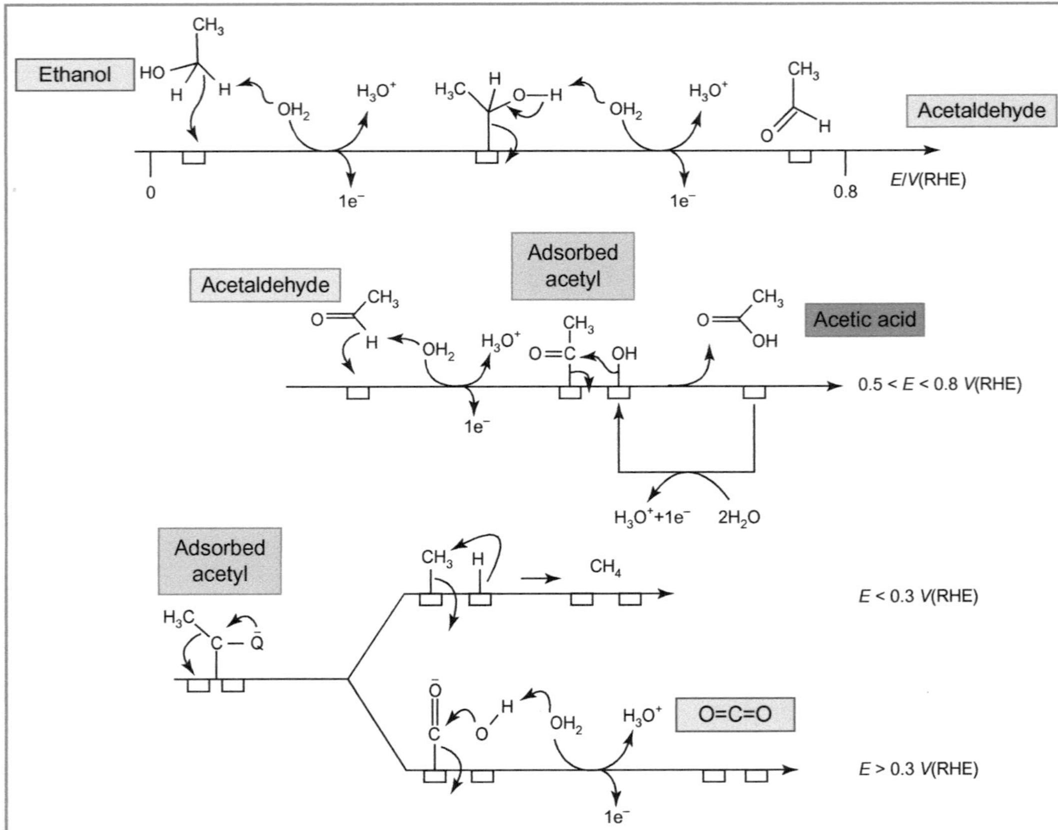

Figure 4.13 Proposed mechanism for the electrocatalytic oxidation of ethanol on a Pt-based electrode in acidic medium.

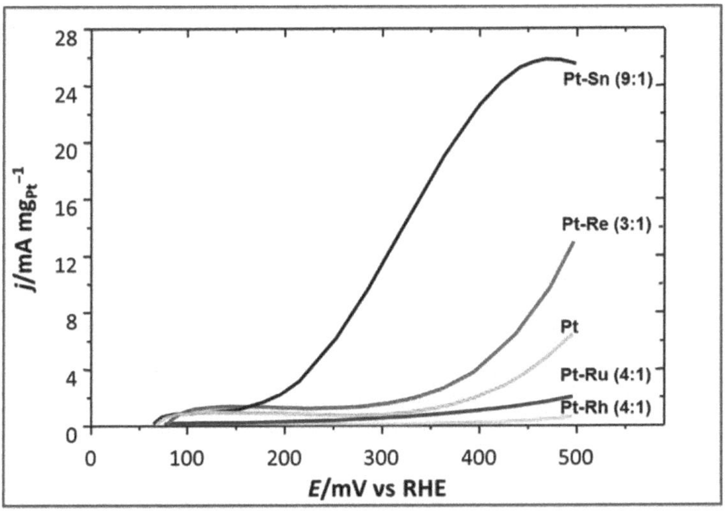

Figure 4.14 Voltammetric curves of the electrooxidation of C_2H_5OH on a RDE covered with different Pt-based bimetallic catalysts deposited on Vulcan XC72 carbon powder. 0.1 M $HClO_4$ + 0.1 M C_2H_5OH; $v = 5$ mV s^{-1}; $\omega = 3000$ rpm; 25°C. *After [51].*

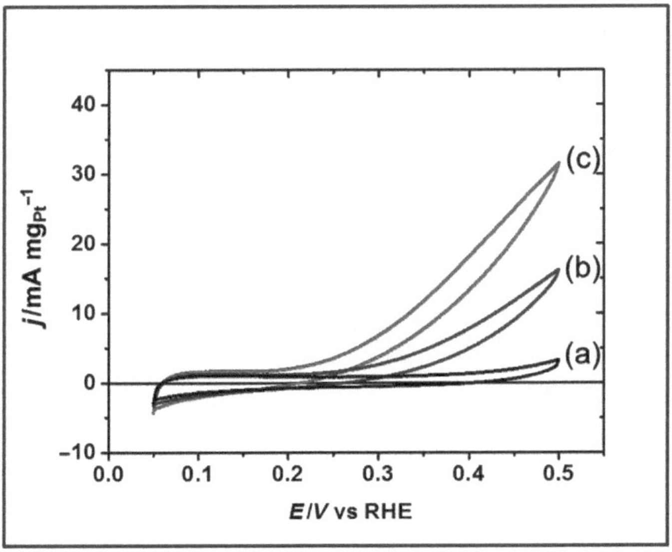

Figure 4.15 Cyclic voltammograms recorded at $v = 5$ mV s^{-1} and $T = 20$°C in a 0.1 M $HClO_4$ + 0.1 M ethanol solution on different Pt-based electrodes with a Pt loading of 0.1 mg$_{Pt}$ cm^{-2}: (a) (Pt/C), (b) $Pt_{0.9}Sn_{0.1}$/C, and (c) $Pt_{0.86}Sn_{0.1}Ru_{0.04}$/C.

previous results [51]. The oxidation of ethanol starts at c.0.30 V versus RHE on pure Pt and c.0.15 V versus RHE on both $Pt_{0.9}Sn_{0.1}/C$ and $Pt_{0.86}Sn_{0.1}Ru_{0.04}/C$ catalysts. However, the ternary catalyst leads to higher current densities over the whole studied potential range, which shows its higher catalytic activity toward ethanol electrooxidation.

4.2.2.3 Principle of ethanol decomposition in a PEMEC

The principle of the electrochemical decomposition of ethanol in a PEMEC is similar to that of water (Fig. 2.1). Similarly, ethanol is fed to the anodic compartment where it can be completely oxidized in the presence of water, producing carbon dioxide and protons, that is:

$$C_2H_5OH + 3\,H_2O \rightarrow 2\,CO_2 + 12\,H^+ + 12\,e^- \qquad (4.10)$$

After crossing-over the proton exchange membrane, the protons reach the cathode compartment where they are reduced to hydrogen according to reaction (2.2).

This corresponds to the electrochemical reforming of ethanol into hydrogen and carbon dioxide, according to the overall reaction (4.11):

$$C_2H_5OH + 3\,H_2O \rightarrow 2\,CO_2 + 6\,H_2 \qquad (4.11)$$

with the thermodynamic data under standard conditions:

$$\Delta H^{+0} = 2\Delta H^{f}_{CO2} - \Delta H^{f}_{C_2H_5OH} - 3\Delta H^{f}_{H_2O} = +\,348\ \text{kJ}\ (\text{mol ethanol})^{-1}$$

$$\text{and } \Delta G^{+0} = 2\Delta G^{f}_{CO2} - \Delta G^{f}{}_{C2H5OH} - 3\Delta G^{f}{}_{H2O}$$
$$= +\,97.3\ \text{kJ}\ (\text{mol ethanol})^{-1}$$

where ΔH^f_i and ΔG^f_i are the thermodynamic data of formation of compound (i) under standard conditions.

Both water electrolysis and ethanol decomposition need external energy ($\Delta H^+ > 0$), coming from the external electrical power source, but in the case of ethanol, the energy needed to produce one hydrogen mole is much smaller than that of water electrolysis: $\Delta H^{+0} = +348/6 = +58$ kJ (mol H_2)$^{-1}$ for ethanol versus $\Delta H^{+0} = +\ 286$ kJ (mol H_2)$^{-1}$ for water decomposition.

The corresponding theoretical cell voltage U^0_{cell}, under standard conditions, can be calculated from ΔG^{+0}, that is, $U^0_{cell} = \Delta G^{+0}/nF$, giving, respectively, $U^0_{H_2O} = 1.229$ V ($n = 2$) for water and $U^0_{EtOH} = 0.084$ V for ethanol electrolysis ($n = 12$). For the DEFC, this corresponds to a standard electromotive force $E^0_{DEFC} = U^0_{H_2O} - U^0_{EtOH} = 1.229 - 0.084 = 1.145$ V.

The theoretical standard cell voltage for ethanol electrochemical reforming ($U^0_{cell} \approx 0.084$ V) is very small compared to that of water electrolysis ($U^0_{cell} \approx 1.23$ V) so that the external energy, proportional to U^0_{cell}—see Eq. (2.11) in Section 2.2, theoretically needed to produce 1 mol of hydrogen by the electrochemical decomposition of ethanol should be much smaller.

However, the relatively slow kinetics of the anodic reaction in both processes, at high current densities, over 1 A cm^{-2}, necessary for high hydrogen production rates, leads to high-cell voltages, greater than 1.8 V for water electrolysis, and at least 0.6 V for ethanol oxidation [50], as illustrated in Fig. 2. From these curves, the cell voltages for water electrolysis (U_{H_2O}), ethanol electrolysis (U_{EtOH}), hydrogen/oxygen fuel cell (E_{FC}), and DEFC (E_{DEFC}) at a current density of 1 A cm^{-2} can be evaluated.

In order to get a competitive cost of energy for the hydrogen production, the cell voltages have to be decreased down to acceptable values, through the development of new electrocatalysts with higher activity and selectivity since the energy consumed is directly proportional to the cell voltage, according to Eq. (2.11)—see Section 2.2. Therefore, U_{cell} must be below 1 V to decrease the energy consumed below 2.2 kWh $(Nm^3)^{-1}$.

4.2.2.4 Electrolyzes of ethanol in a PEMEC

Electrolyses of 2 M ethanol were carried out in a PEMEC at 20°C and at two constant controlled current densities ($j = 50$ and 100 mA cm^{-2}) with the three Pt-based electrocatalysts: Pt/C, $Pt_{0.9}Sn_{0.1}$/C, and $Pt_{0.86}Sn_{0.10}Ru_{0.04}$/C. In each case, the cell voltage U_{cell} was recorded as a function of time ($t = 0-30$ minutes), and the evolved hydrogen was measured with a gas flow meter. In contrast, the analysis of the reaction products was carried out by high-performance liquid chromatography (HPLC) after 30 minutes of electrolysis at 100 mA cm^{-2}.

The $U_{cell}(t)$ curves are given in Fig. 4.16 at 100 mA cm^{-2} for PtC and 50 and 100 mA cm^{-2} for the binary and ternary catalysts. It has to be stated that in the case of pure Pt, very important potential instabilities appeared for the experiment at 50 mA cm^{-2}, whereas in the case of the binary and ternary catalysts, short-period potential oscillations were observed (not shown here). For all catalysts, the cell voltage was always higher after 30 minutes of electrolysis than the initial values,

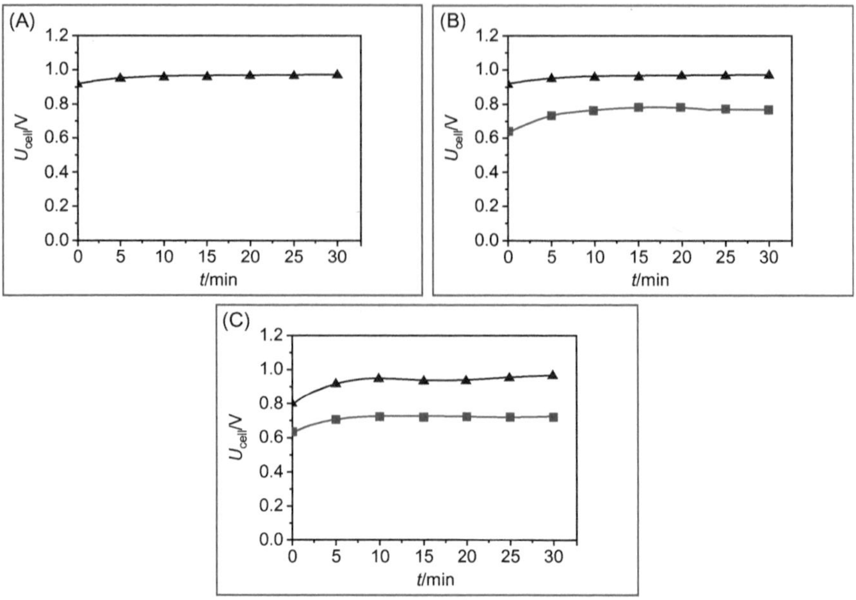

Figure 4.16 Electrolysis cell voltage versus time, U_{cell} (t), at 20°C for a PEMEC fitted with a Pt/C cathode and (A) a Pt/C anode, (B) a $Pt_{0.9}Sn_{0.1}/C$ anode, and (C) a $Pt_{0.86}Sn_{0.10}Ru_{0.04}/C$ anode. 2 M ethanol, $V_{initial} = 1$ L, flow rate = 2 mL $minute^{-1}$; (▲) $j = 100$ mA cm^{-2}; and (■) $j = 50$ mA cm^{-2}.

independently of the current density used. This deactivation was due the formation of strongly adsorbed intermediates occurring during the electro-oxidation of ethanol, and it was possible to reactivate the catalysts by increasing the cell voltage up to 1.0 V.

Under such anode potential conditions, previously adsorbed CO from the dissociative adsorption of ethanol is oxidized, and at the same time, either no adsorbed CO is formed from ethanol adsorption, or the adsorbed CO oxidation kinetics is sufficiently fast to avoid its accumulation at the electrode surface and the blockage of the catalytic sites. Therefore, such treatment makes the catalyst surface clean. One can notice that this explanation is in agreement with the constant value of U_{cell} observed at c.1 V for the 100 mA cm^{-2} electrolysis with all catalysts, whereas some potential oscillations were observed for a lower current density, for example, 50 mA cm^{-2}, between 0.6 and 0.75 V. Such oscillations were already observed for the oxidation of methanol, formic acid, formaldehyde, ethanol, propanol, and other organic compounds at platinum-based electrodes in acidic as well as in alkaline media [69–75].

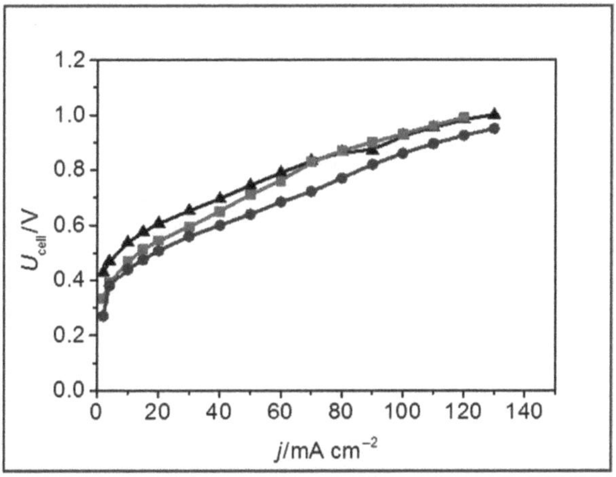

Figure 4.17 Electrolysis cell voltage versus current density, $U_{cell}(j)$, for the oxidation of 2 M CH_3CH_2OH in the PEMEC (0.5 M H_2SO_4, Pt/C, N117, $Pt_xSn_yRu_z$/C, $V_{initial} = 1$ L, flow rate = 2 mL minute^{-1}) at 20°C. (▲) Pt/C anode, (■) $Pt_{0.9}Sn_{0.1}$/C anode, and (●) $Pt_{0.86}Sn_{0.10}Ru_{0.04}$/C.

The occurrence of these oscillations was explained in terms of cycles of accumulation and consumption of adsorbed species, such as poisoning species, anions, oxygenated species, leading to instabilities under the steady-state situation [76].

Fig. 4.17 shows the polarization curves recorded for an electrolysis cell fitted with the different anode catalysts. The cell voltage U_{cell} versus current density j curves displays the expected logarithmic behavior. The same behavior is observed in the electrolysis cell as in a classical three-electrode electrochemical cell, for example, both the binary and ternary catalysts lead to similar cell voltages at a given current density, lower than that achieved with the pure Pt catalyst in the low potential region. However, in the high potential region, where the added elements (Sn, Ru) are no more stable and can dissolve, the $U_{cell}(j)$ curve of the $Pt_{0.9}Sn_{0.1}$/C catalyst is quite similar to that of the Pt/C catalyst.

4.2.2.5 Measurement of the hydrogen evolution rate

The hydrogen evolution rate was determined by measuring the volume of evolved hydrogen as a function of time for three different current densities (j = 20, 50, and 100 mA cm^{-2}) during the electrolysis of a 2 M C_2H_5OH solution (Fig. 4.18).

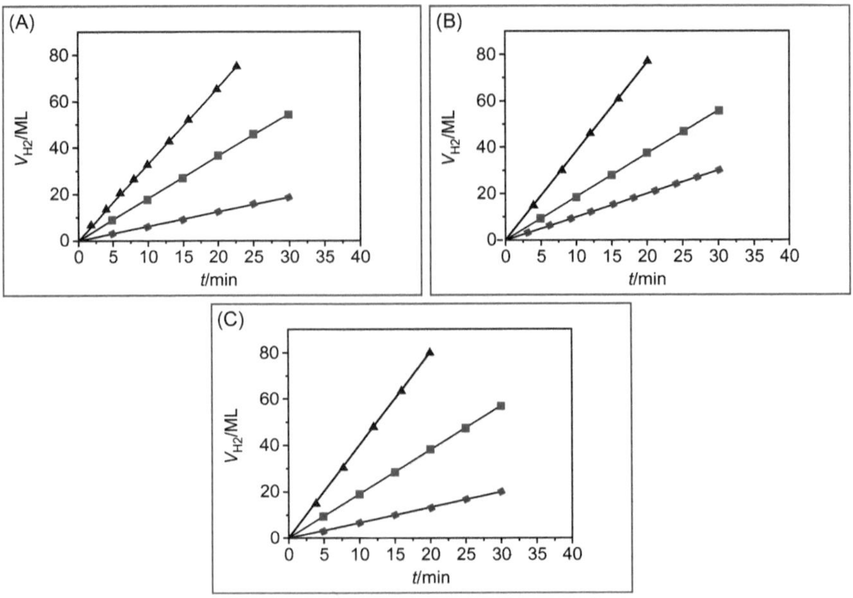

Figure 4.18 Hydrogen evolution at different current densities for the PEMEC (0.5 M H_2SO_4, Pt/C, N117, (A) Pt/C, (B) $Pt_{0.9}Sn_{0.1}/C$, (C) $Pt_{0.86}Sn_{0.10}Ru_{0.04}/C$, 2 M CH_3CH_2OH) ($V_{initial} = 1$ L, flow rate = 2 mL minute^{-1}) at 20°C. (◆) $j = 20$ mA cm^{-2}, (●) $j = 50$ mA cm^{-2}, and (▲) $j = 100$ mA cm^{-2}.

Table 4.6 Comparison of the experimental volume of hydrogen (V_{H2}), obtained after 30 min for the electrolysis of 2 M ethanol at 100 mA cm^{-2}, with the theoretical value.

Electrocatalyst	V_{H2} experimental/mL (after 30 min at 0.5 A)	V_{H2} theoretical/mL at 20°C (after 30 min at 0.5 A)
Pt/C	102	112
PtSn/C	111	112
PtSnRu/C	108	112

In all experiments, the measured volume of hydrogen at a given current density, $j = 20$, 50, and 100 mA cm^{-2}, is a linear function of time with correlation coefficients > 0.9968, showing clearly that the volume of evolved hydrogen does not depend on the C_2H_5OH concentration, nor on the nature of the anode catalyst, but only on the quantity of electricity involved, that is, $\Delta q = I\ \Delta t$ where Δt is the electrolysis time, according to the Faraday law—see Eq. (2.13) in Section 2.2.

Table 4.6 summarizes all the results obtained with the three electrocatalysts, for example, Pt/C, $Pt_{0.9}Sn_{0.1}/C$, and $Pt_{0.86}Sn_{0.1}Ru_{0.04}/C$, investigated here, where V_{H2} experimental is the volume of evolved hydrogen after 30 minutes electrolysis of 2 M ethanol at 100 mA cm^{-2}, that is, at 0.5 A. These experimental results are compared to the calculated values of the volume of hydrogen produced after 30 minutes of electrolysis according to the Faraday's law—see Eq. (2.13) in Section 2.2. The agreement is very good in all experiments, but with experimental values slightly lower than those calculated, which may come from some gas leakage, either in the experimental setup, or through the proton exchange membrane.

For all experiments, the electrical energy needed to produce 1 Nm^3 of hydrogen was also evaluated, since it is only a function of the cell voltage U_{cell}, according to Eq. (2.11) in Section 2.2. As an example, the values obtained for U_{cell}, during ethanol electrolysis at a $Pt_{0.9}Sn_{0.10}/C$ anode from 0 to 30 minutes, are given in Table 4.7, together with the electrical energy used to produce 1 Nm^3 of hydrogen.

In all the results obtained, the amount of electrical energy is below 2.2 kWh $(Nm^3)^{-1}$, since $U_{cell} < 1.0$ V, which is at least 2—2.5 times lower than the energy consumed for water electrolysis.

4.2.2.6 Analysis of the reaction products of ethanol oxidation

In order to determine the reaction products, the current density was kept constant, and the voltage of the cell was measured as a function of time. The ethanol flow rate was chosen at 2 mL $minute^{-1}$ in order to perform the experiments for 30 minutes. Because the electrodes are homemade, fluctuations in the amount of catalyst deposited on the electrode can

Table 4.7 Cell voltage, electrical energy, experimental, and theoretical volume of generated H_2 as a function of time for the electrolysis of 2 M ethanol at 20°C in a 5 cm^2 PEMEC under the same conditions as in Fig. 4.16 (B) for $Pt_{0.9}Sn_{0.10}/C$ (100 mA cm^{-2}).

Time/s	Cell voltage/V	Electrical energy/ kWh $(Nm^3)^{-1}$	Volume of evolved H_2/mL	Theoretical volume of H_2 at 20°C/mL
0	0.920	2.02	0	0
300	0.951	2.08	18	18.7
600	0.958	2.09	35	37.4
900	0.960	2.10	54	56.1
1200	0.962	2.11	72	74.8
1500	0.968	2.12	91.5	93.5
1800	0.972	2.13	111	112

occur. Then, in order to compare the behavior of the different catalysts, for example, Pt/C, $Pt_{0.9}Sn_{0.1}/C$, and $Pt_{0.86}Sn_{0.10}Ru_{0.04}/C$, in terms of selectivity, it is preferable to express the selectivity as a nondimensional parameter such as the ratio between the number of moles of a produced species to the total number of moles of all the formed products, which corresponds rather to the chemical yield for each reaction product.

Table 4.8 reports the results obtained at a current density of 100 mA cm^{-2}, that is, a current intensity $I = 0.5$ A, for the three different catalysts. With all catalysts, only AAL and AA were detected by chromatographic analyses. Thus, the amount of CO_2 formed was estimated by comparing the quantity of electricity involved in the formation of AAL and AA to the whole quantity of electricity, Δq, involved after 20 minutes of electrolysis:

$$\Delta q = I\,\Delta t = 0.5 \times 30 \times 60 = 900\ \text{C}$$

From the results given in Table 4.8, it appears that the addition of tin to platinum greatly favors the cleavage of the C–C bond at such high anode potentials, since the yield in CO_2 becomes more than twice higher with the $Pt_{0.9}Sn_{0.1}/C$ and the $Pt_{0.86}Sn_{0.10}Ru_{0.04}/C$ catalysts than with the Pt/C catalyst.

The evaluation of the total number of Faradays per mole of ethanol exchanged during the oxidation process, that is, $n_e \approx 3.5-4.6$, suggests that the main reaction product is AA ($n_e = 4$). Thus, assuming that the mean number of Faraday per mole of ethanol is about 4, as determined from the quantitative analysis of reaction products, that is, that the oxidation of ethanol stops at AA, the electrochemical reactions involved in the electrolysis process are as follows:

$$CH_3CH_2OH + H_2O \rightarrow CH_3COOH + 4\,H^+ + 4\,e^- \quad (4.8)$$

$$\text{and} \quad 4H^+ + 4\,e^- \rightarrow 2\,H_2 \quad (2.2)$$

corresponding to the overall reaction:

$$CH_3CH_2OH + H_2O \rightarrow CH_3COOH + 2\,H_2 \qquad (4.12)$$

with the following thermodynamic data under standard conditions:

$$\Delta H^{+0} = +\,79.1\ \text{kJ/mol}; \quad \Delta G^{+0} = +\,22.1\ \text{kJ/mol}$$

The number of moles of ethanol transformed after 30 minutes of electrolysis can then be evaluated, assuming $n_e \approx 4$, as follows:

Table 4.8 Chemical yield and number of electrons exchanged per mole of ethanol for the electrolysis of 2 M ethanol at 0.5 A during 30 min ($j = 100$ mA cm^{-2}).

Catalyst		AAL	AA	CO_2	Ethanol consumed/mole	Calculated n_e[a]
Pt/C	Solution volume 0.75 L		$U_{cell} \approx 0.92–0.97$ V		$2.62\ 10^{-3}$	3.5
	Amount of substance/mol	$1.28\ 10^{-3}$	$1.16\ 10^{-3}$	$3.54\ 10^{-4}$		
	Chemical yield/mole%	**46**	**42**	**12**		
$Pt_{0.9}Sn_{0.1}/C$	Solution volume 0.29 L		$U_{cell} \approx 0.92 - 0.97$ V		$2.02\ 10^{-3}$	4.6
	Amount of substance/mol	$7.17\ 10^{-4}$	$9.63\ 10^{-4}$	$6.74\ 10^{-4}$		
	Chemical yield/mole%	**30**	**41**	**29**		
$Pt_{0.86}Sn_{0.1}Ru_{0.04}/C$	Solution volume 0.24 L		$U_{cell} \approx 0.80 - 0.96$ V		$2.09\ 10^{-3}$	4.5
	Amount of substance/mol	$9.52\ 10^{-4}$	$7.76\ 10^{-4}$	$7.20\ 10^{-4}$		
	Chemical yield/mole%	**39**	**32**	**29**		

[a]Total number of mole of electrons exchanged for the oxidation of 1 mol of ethanol, assuming that no other electrolysis products than AAL, AA, and CO_2 are formed.

$$\Delta N = \Delta q / n_e F = 900/(4 \times 96485) = 2.33\ 10^{-3}\ \text{mol}$$

which corresponds to a very small fraction of ethanol converted and to a low reaction rate:

$$v = 2.33\ 10^{-3}/1800 = 1.3\ 10^{-6}\ \text{mol}^{-1} = 4.66\ 10^{-3}\ \text{mol h}^{-1}$$

In contrast, it is possible to evaluate the energy efficiency of the process (see Eq. 2.15 (a) in Section 2.4), either for a complete oxidation of ethanol into CO_2 involving 12 Faraday mol^{-1} or for a partial oxidation stopping at the AA stage involvig 4 Faraday mol^{-1}.

In the first case, the energy efficincy $\varepsilon_{\text{EtOH} \rightarrow \text{CO2}}$ can be evaluated at 20°C for the complete oxidation of ethanol on the $Pt_{0.9}Sn_{0.1}/C$ electrocatalyst working at 100 mA cm^{-2} and at a cell voltage $U_{\text{cell}} = 0.972$ V (see Table 4.7):

$$\varepsilon_{\text{EtOH} \rightarrow \text{CO}_2} = 348/(348 + 1028) = 348/1376 = 0.253 = 25.3\%$$

with an energy loss $nF\ \eta_{\text{loss}} = 12 \times 96.485 \times 0.972 - 97.3 = 1028$ kJ.

In the second case, corresponding to the formation of AA at the same cell voltage $U_{\text{cell}} = 0.972$ V, the energy efficiency would be:

$$\varepsilon_{\text{EtOH} \rightarrow \text{AA}} = 79.1/(79.1 + 353) = 79.1/432.1 = 0.183 = 18.3\%$$

with an energy loss $nF\ \eta_{\text{loss}} = 4 \times 96.485 \times 0.972 - 22.1 = 353$ kJ.

These energy efficiencies are relatively small compared with that of methanol electrochemical reforming, that is, $\varepsilon_{\text{cell}} \approx 35\%$ at 70°C, and above all to that of water electrolysis at 1.8 V and 25°C, that is, $\varepsilon_{\text{cell}} \approx 70\%$.

References

[1] A. Capon, R. Parsons, The oxidation of formic acid on noble metal electrodes: II. A comparison of the behaviour of pure electrodes, J. Electroanal. Chem. 44 (1973) 239–254. Available from: https://doi.org/10.1016/S0022-0728(73)80250-5.

[2] X. Xia, T.J. Iwasita, Influence of underpotential deposited lead upon the oxidation of HCOOH in $HClO_4$ at platinum electrodes, J. Electrochem. Soc. 140 (1993) 2559–2565. Available from: https://doi.org/10.1149/1.2220862.

[3] N. Markovic, H. Gaseiger, P. Ross, X. Jian, I. Villegas, M. Weaver, Electro-oxidation mechanisms of methanol and formic acid on Pt-Ru alloy surfaces, Electrochim. Acta 40 (1995) 91–98. Available from: https://doi.org/10.1016/0013-4686(94)00241-R.

[4] T.D. Jarvi, E.M. Stuve, Fundamental aspects of vacuum and electrocatalytic reactions of methanol and formic acid on platinum surfaces, in: J. Lipkowski, P.N. Ross (Eds.), Electrocatalysis, Wiley, New York, 1998, pp. 75–154. ISBN: 0-47-24673-5.

[5] P.N. Ross, The science of electrocatalysis on bimetallic surfaces, in: J. Lipkowski, P.N. Ross (Eds.), Electrocatalysis, Wiley, New York, 1998, pp. 63–74. ISBN: 0-47-24673-5.

[6] G.Q. Lu, A. Crown, A. Wieckowski, Formic acid decomposition on polycrystalline platinum and palladized platinum electrodes, J. Phys. Chem. B 103 (1999) 9700–9711. Available from: https://doi.org/10.1021/jp992297x.
[7] X. Yu, P.G. Pickup, Recent advances in direct formic acid fuel cells (DFAFC), J. Power Sources 182 (2008) 124–132. Available from: https://doi.org/10.1016/j.jpowsour.2008.03.075.
[8] H. Meng, D. Zeng, F. Xie, Recent development of Pd-based electrocatalysts for proton exchange membrane fuel cells, Catalysts 5 (2015) 1221–1274. Available from: https://doi.org/10.3390/catal5031221.
[9] E.Ö. Kiliç, A.S. Koparal, Ü.B. Öğütveren, Hydrogen production by electrochemical decomposition of formic acid via solid polymer electrolyte, Fuel Proc. Technol. 90 (2009) 158–163. Available from: https://doi.org/10.1016/j.fuproc.2008.08.011.
[10] L. Aldous, R.G. Compton, Clean, efficient electrolysis of formic acid via formation of eutectic, ionic mixtures with ammonium formate, Energy Environ. Sci. 3 (2010) 1587–1592. Available from: https://doi.org/10.1039/C0EE00151A.
[11] C. Lamy, A. Devadas, M. Simoes, C. Coutanceau, Clean hydrogen production through the electrocatalytic oxidation of formic acid in a PEM electrolysis cell (PEMEC), Electrochim. Acta 60 (2012) 112–120. Available from: https://doi.org/10.1016/j.electacta.2011.11.006.
[12] R. Larsen, S. Ha, J. Zakzeski, R.I. Masel, Unusually active palladium-based catalysts for the electrooxidation of formic acid, J. Power Sources 157 (2006) 78–84. Available from: https://doi.org/10.1016/j.jpowsour.2005.07.066.
[13] V. Mazumder, S. Sun, Oleylamine-mediated synthesis of Pd nanoparticles for catalytic formic acid oxidation, J. Am. Chem. Soc. 131 (2009) 4588–4589. Available from: https://doi.org/10.1021/ja9004915.
[14] W. Zhou, J.Y. Lee, Highly active core–shell Au@Pd catalyst for formic acid electrooxidation, Electrochem. Commun. 9 (2007) 1725–1729. Available from: https://doi.org/10.1016/j.elecom.2007.03.016.
[15] C. Rice, S. Ha, R.I. Masel, A. Wieckowski, Catalysts for direct formic acid fuel cells, J. Power Sources 115 (2003) 229–235. Available from: https://doi.org/10.1016/S0378-7753(03)00026-0.
[16] G. Zhang, Y. Wang, X. Wang, Y. Chen, Y. Zhou, Y. Tang, et al., Preparation of Pd–Au/C catalysts with different alloying degree and their electrocatalytic performance for formic acid oxidation, Appl. Catal. B: Environ 102 (2011) 614–619. Available from: https://doi.org/10.1016/j.apcatb.2010.12.049.
[17] A. Bewick, B. Beden, C. Lamy, Oscillatory kinetics in the electrochemical oxidation of formate ions during the deposition of rhodium electrode: part II. Mechanistic considerations, J. Electroanal. Chem. 121 (1981) 115–124. Available from: https://doi.org/10.1016/S0022-0728(81)80572-4.
[18] B.I. Podlovchenko, O.A. Petry, A.N. Frumkin, H. Lal, The behavior of a platinized-platinum electrode in solutions of alcohols containing more than one carbon atom, aldehydes and formic acid, J. Electroanal. Chem. 11 (1966) 12–25. Available from: https://doi.org/10.1016/0022-0728(66)80053-0.
[19] V.S. Bagotzky, Y.B. Vassiliev, O.A. Khazova, Generalized scheme of chemisorption, electrooxidation and electroreduction of simple organic compounds on platinum group metals, J. Electroanal. Chem. 81 (1977) 229–238. Available from: https://doi.org/10.1016/S0022-0728(77)80019-3.
[20] H.R. Corti, E.R. Gonzalez, in: H.R. Corti, E.R. Gonzalez (Eds.), Direct alcohol fuel cells, materials, performance, durability and applications, Springer, Dordrecht/Heidelberg/New York/London, 2014, ISBN: 978-94-007-7708-8, .
[21] M. Bellini, M. Bevilacqua, M. Innocenti, A. Lavachi, H.A. Miller, J. Filippi, et al., Energy & chemicals from renewable resources by electrocatalysis, J. Electrochem.

Soc. 161 (2014) D3032–D3043. Available from: https://doi.org/10.1149/2.005407jes.
[22] C. Coutanceau, S. Baranton, Electrochemical conversion of alcohols for hydrogen production: a short review, Wire Energy Environ. 5 (2016) 388–400. Available from: https://doi.org/10.1002/wene.193.
[23] F.M. Sapountzi, M.N. Tsampas, H.O.A. Fredriksson, J.M. Gracia, J.W. Niemantsverdriet, Hydrogen from electrochemical reforming of C1-C3 alcohols using proton conducting membranes, Int. J. Hydrogen Energy 42 (2017) 10762–10774. Available from: https://doi.org/10.1016/j.ijhydene.2017.02.195.
[24] H.K. Ju, S. Giddey, S.P.S. Badwal, R.J. Mulder, Electro-catalytic conversion of ethanol in solid electrolyte cells for distributed hydrogen generation, Electrochim. Acta 212 (2016) 744–757. Available from: https://doi.org/10.1016/j.electacta.2016.07.062.
[25] T. Take, K. Tsurutani, M. Umeda, Hydrogen production by methanol–water solution electrolysis, J. Power Sources 164 (2007) 9–16. Available from: https://doi.org/10.1016/j.jpowsour.2006.10.011.
[26] S. Uhm, H. Jeon, T.J. Kim, J. Lee, Clean hydrogen production from methanol–water solutions via power saved electrolytic reforming process, J. Power Sources 198 (2012) 218–222. Available from: https://doi.org/10.1016/j.jpowsour.2011.09.083.
[27] Z. Hu, M. Wu, Z. Wei, S. Song, P.K. Shen, Pt-WC/C as a cathode electrocatalyst for hydrogen production by methanol electrolysis, J. Power Sources 166 (2007) 458–461. Available from: https://doi.org/10.1016/j.jpowsour.2007.01.083.
[28] S.R. Narayanan, W. Chun, B. Jeffries-Nakamura, T.I. Valdez. Hydrogen generation by electrolysis of aqueous organic solutions, US Patent (2006) 7056428 B2.
[29] G. Sasikumar, A. Muthumeenal, S.S. Pethaiah, N. Nachiappan, R. Balaji, Aqueous methanol electrolysis using proton conducting membrane for hydrogen production, Int. J. Hydrogen Energy 33 (2008) 5905–5910. Available from: https://doi.org/10.1016/j.ijhydene.2008.07.013.
[30] C.R. Cloutier, D.P. Wilkinson, Electrolytic production of hydrogen from aqueous acidic methanol solutions, Int. J. Hydrogen Energy 35 (2010) 3967–3984. Available from: https://doi.org/10.1016/j.ijhydene.2010.02.005.
[31] C. Lamy, B. Guenot, M. Cretin, G. Pourcelly, Kinetics analysis of the electrocatalytic oxidation of methanol inside a DMFC working as a PEM electrolysis cell (PEMEC) to generate clean hydrogen, Electrochim. Acta 177 (2015) 352–358. Available from: https://doi.org/10.1016/j.electacta.2015.02.069.
[32] C. Lamy, Anodic reactions in electrocatalysis: methanol oxidation, in: G. Kreysa, K.I. Ota, R.F. Savinell (Eds.), Encyclopedia of Applied Electrochemistry, Springer Online, 2014, pp. 85–92. Available from: https://doi.org/10.1007/978-1-4419-6996-5_405.
[33] J.C. Amphlett, K.A.M. Creber, J.M. Davis, R.F. Mann, B.A. Peppley, D.M. Stokes, Hydrogen production by steam reforming of methanol for polymer electrolyte fuel cells, Int. J. Hydrogen Energy 19 (1994) 131–137. Available from: https://doi.org/10.1016/0360-3199(94)90117-1.
[34] S. Sá, H. Silva, L. Brandão, J.M. Sousa, A. Mendes, Catalysts for methanol steam reforming-A review, Appl. Catal. B: Environ. 99 (2010) 43–57. Available from: https://doi.org/10.1016/j.apcatb.2010.06.015.
[35] Y. Matsumura, Durable Cu composite catalyst for hydrogen production by high temperature methanol steam reforming, J. Power Sources 272 (2014) 961–969. Available from: https://doi.org/10.1016/j.jpowsour.2014.09.047.
[36] A. Kabbabi, R. Faure, R. Durand, B. Beden, F. Hahn, J.M. Léger, et al., In situ FTIRS study of the electrocatalytic oxidation of carbon monoxide and methanol at

platinum-ruthenium bulk alloy electrodes, J. Electroanal. Chem. 444 (1998) 41–53. Available from: https://doi.org/10.1016/S0022-0728(97)00558-5.
[37] C. Lamy, A. Lima, V. LeRhun, F. Delime, C. Coutanceau, J.M. Léger, Recent advances in the development of direct alcohol fuel cells (DAFC), J. Power Sources 105 (2002) 283–296. Available from: https://doi.org/10.1016/S0378-7753(01)00954-5.
[38] M. Watanabe, S. Motoo, Electrocatalysis by ad-atoms. 2. Enhancement of oxidation of methanol on platinum by ruthenium ad-atoms, J. Electroanal. Chem. 60 (1975) 267–273. Available from: https://doi.org/10.1016/S0022-0728(75)80261-0.
[39] C. Lamy, J.M. Léger, S. Srinivasan, Direct methanol fuel cells: From a twentieth century electrochemist's dream to a twenty-first century emerging technology, in: J. O.'M. Bockris, B.E. Conway, R.E. White (Eds.), Modern Aspects of Electrochemistry, vol. 34, Kluwer Academic/Plenum Publishers, New York, 2001, pp. 53–118. Available from: https://doi.org/10.1007/0-306-46923-5_3.
[40] E.M. Belgsir, H. Huser, J.M. Léger, C. Lamy, A kinetic analysis of the oxidation of methanol at platinum-based electrodes by quantitative determination of the reaction products using liquid chromatography, J. Electroanal. Chem. 225 (1994) 281–286. Available from: https://doi.org/10.1016/0022-0728(87)80022-0.
[41] N.H. Li, S.G. Sun, S.P. Chen, Studies on the role of oxidation states of the platinum surface in electrocatalytic oxidation of small primary alcohols, J. Electroanal. Chem. 430 (1997) 57–67. Available from: https://doi.org/10.1016/S0022-0728(97)00094-6.
[42] B. Guenot, M. Cretin, C. Lamy, Clean hydrogen generation from the electrocatalytic oxidation of methanol inside a proton exchange membrane electrolysis cell (PEMEC): effect of methanol concentration and working temperature, J. Appl. Electrochem. 45 (2015) 973–981. Available from: https://doi.org/10.1007/s10800-015-0867-3.
[43] L. Dubau, C. Coutanceau, E. Garnier, J.M. Léger, C. Lamy, Electrooxidation of methanol at platinum-ruthenium catalysts prepared from colloidal precursors: atomic composition and temperature effect, J. Appl. Electrochem. 33 (2003) 419–429. Available from: https://doi.org/10.1023/A:1024491007321.
[44] A. Lima, C. Coutanceau, J.M. Léger, C. Lamy, Investigation of ternary catalysts for methanol electrooxidation, J. Appl. Electrochem. 31 (2001) 379–386. Available from: https://doi.org/10.1023/A:101757.
[45] C. Lamy, From hydrogen production by water electrolysis to its utilization in a PEM fuel cell or in a solid oxide fuel cell: some considerations on the energy efficiencies, Int. J. Hydrogen Energy 41 (2016) 15415–15425. Available from: https://doi.org/10.1016/j.ijhydene.2016.04.173.
[46] C. Lamy, C. Coutanceau, J.M. Léger, The direct ethanol fuel cell: a challenge to convert bioethanol cleanly into electric energy, in: P. Barbaro, C. Bianchini (Eds.), Catalysis for Sustainable Energy Production, Wiley-VCH, Weinheim, 2009, pp. 3–46. Available from: https://doi.org/10.1002/9783527625413.ch1.
[47] S.P.S. Badwal, S. Giddey, A. Kulkarni, J. Goel, S. Basu, Direct ethanol fuel cells for transport and stationary applications – a comprehensive review, Appl. Energy 145 (2015) 80–103. Available from: https://doi.org/10.1016/j.apenergy.2015.02.002.
[48] A. Caravaca, F.M. Sapountzi, A. De Lucas-Consuegra, C. Molina-Mora, F. Dorado, J.L. Valverde, Electrochemical reforming of ethanol-water solutions for pure H_2 production in a PEM electrolysis cell, Int. J. Hydrogen Energy 37 (2012) 9504–9513. Available from: https://doi.org/10.1016/j.ijhydene.2012.03.062.
[49] A. Caravaca, A. De Lucas-Consuegra, A.B. Calcerrada, J. Lobato, J.L. Valverde, F. Dorado, From biomass to pure hydrogen: electrochemical reforming of bio-ethanol

in a PEM electrolyzer, Appl. Catal. B: Environ. 134–135 (2013) 302–309. Available from: https://doi.org/10.1016/j.apcatb.2013.

[50] S. Rousseau, C. Coutanceau, C. Lamy, J.M. Léger, Direct ethanol fuel cell (DEFC): electrical performances and reaction products distribution under operating conditions with different platinum-based anodes, J. Power Sources 158 (2006) 18–24. Available from: https://doi.org/10.1016/j.jpowsour.2005.08.027.

[51] C. Lamy, S. Rousseau, E.M. Belgsir, C. Coutanceau, J.M. Léger, Recent progress in the direct ethanol fuel cell: development of new platinum–tin electrocatalysts, Electrochim. Acta 49 (2004) 3901–3908. Available from: https://doi.org/10.1016/j.electacta.2004.01.078.

[52] H. Hitmi, E.M. Belgsir, J.M. Léger, C. Lamy, R.O. Lezna, A kinetic analysis of the electrooxidation of ethanol at a platinum electrode in acid medium, Electrochim. Acta 39 (1994) 407–415. Available from: https://doi.org/10.1016/0013-4686(94)80080-4.

[53] T. Iwasita, E. Pastor, A DEMS and FTIR spectroscopic investigation of adsorbed ethanol on polycrystalline platinum, Electrochim. Acta 39 (1994) 531–537. Available from: https://doi.org/10.1016/0013-4686(94)80097-9.

[54] T. Iwasita, E. Pastor, D/H exchange of ethanol at platinum-electrodes, Electrochim. Acta 39 (1994) 547–551. Available from: https://doi.org/10.1016/0013-4686(94)80099-5.

[55] J. Wang, S. Wasmus, R.F. Savinell, Evaluation of ethanol, 1-propanol, and 5-propanol in a direct oxidation polymer-electrolyte fuel cell - A real-time mass spectrometry study, J. Electrochem. Soc. 142 (1995) 4218–4224. Available from: https://doi.org/10.1149/1.2048487.

[56] J. Souza, F.J.B. Rabelo, I.R. De Moraes, F.C. Nart, Performance of a co-electrodeposited Pt-Ru electrode for the electro-oxidation of ethanol studied by in situ FTIR spectroscopy, J. Electroanal. Chem. 420 (1997) 17–20. Available from: https://doi.org/10.1016/S0022-0728(96)01019-4.

[57] F. Delime, J.M. Léger, C. Lamy, Enhancement of the electrooxidation of ethanol on a Pt-PEM electrode modified by tin. Part I: half-cell study, J. Appl. Electrochem. 29 (1999) 1249–1254. Available from: https://doi.org/10.1023/A:100378.

[58] J.M. Perez, B. Beden, F. Hahn, A. Aldaz, C. Lamy, In situ infrared reflectance spectroscopic study of the early stages of ethanol adsorption at a platinum electrode in acid medium, J. Electroanal. Chem. 262 (1989) 251–261. Available from: https://doi.org/10.1016/0022-0728(89)80026-9.

[59] H. Wang, Z. Jusys, R.J. Behm, Ethanol electro-oxidation on carbon-supported Pt, PtRu and Pt_3Sn catalysts: a quantitative DEMS study, J. Power Sources 154 (2006) 351–359. Available from: https://doi.org/10.1016/j.jpowsour.2005.10.034.

[60] M. Watanabe, S. Motoo, Electrocatalysis by ad-atoms: Part III. Enhancement of the oxidation of carbon monoxide on platinum by ruthenium ad-atoms, J. Electroanal. Chem. 60 (1975) 275–283. Available from: https://doi.org/10.1016/S0022-0728(75)80262-2.

[61] C. Lamy, E.M. Belgsir, Other direct alcohol fuel cells, in: W. Vielstich, A. Lamm, H. Gasteiger (Eds.), Handbook of fuel cells: fundamentals and survey of systems, vol. 1, Wiley, Chichester, 2003, pp. 323–334. Available from: https://doi.org/10.1002/9780470974001.f104018.

[62] C. Lamy, Direct alcohol fuel cells (DAFCs), in: G. Kreysa, K.I. Ota, R.F. Savinell (Eds.), Encyclopedia of applied electrochemistry, Springer Online, 2014, ISBN: 978-1-4419-6996-5, , pp. 321–330. Online. Available from: https://doi.org/10.1007/978-1-4419-6996-5_188.

[63] C. Lamy, T. Jaubert, S. Baranton, C. Coutanceau, Clean hydrogen generation through the electrocatalytic oxidation of ethanol in a proton exchange membrane

electrolysis cell (PEMEC). Effect of the nature and structure of the catalytic anode, J. Power Sources 245 (2014) 927–936. Available from: https://doi.org/10.1016/j.jpowsour.2013.07.028.

[64] F. Vigier, C. Coutanceau, A. Perrard, E.M. Belgsir, C. Lamy, Development of anode catalysts for a direct ethanol fuel cell, J. Appl. Electrochem. 34 (2004) 439–446. Available from: https://doi.org/10.1023/B:JACH.0000016629.98535.ad.

[65] N.R. De Tacconi, R.O. Lezna, B. Beden, F. Hahn, C. Lamy, In-situ FTIR study of the electrocatalytic oxidation of ethanol at iridium and rhodium electrodes, J. Electroanal. Chem. 379 (1994) 329–337. Available from: https://doi.org/10.1016/0022-0728(94)87155-8.

[66] E. Mendez, J.L. Rodriguez, M.C. Arevalo, E. Pastor, Comparative study of ethanol and acetaldehyde reactivities on rhodium electrodes in acidic media, Langmuir 18 (2002) 763–772. Available from: https://doi.org/10.1021/la010747i.

[67] A. Bonesi, G. Garaventa, W. Triaca, A.M. Castro Luna, Synthesis and characterization of new electrocatalysts for ethanol oxidation, Int. J. Hydrogen Energy 33 (2008) 3499–3501. Available from: https://doi.org/10.1016/j.ijhydene.2007.12.056.

[68] F. Colmati, E. Antolini, E.R. Gonzalez, Preparation, structural characterization and activity for ethanol oxidation of carbon supported ternary Pt–Sn–Rh catalysts, J. Alloys Comp. 456 (2008) 264–270. Available from: https://doi.org/10.1016/j.jallcom.2007.02.015.

[69] J. Wojtowicz, Oscillatory behavior in electrochemical systems, in: J.O.'M. Bokris, B.E. Conway (Eds.), Modern Aspects of Electrochemistry, vol. 8, Plenum Press, New York, 1973, , ISBN: 978-1-4615-7440-8, pp. 47–120.

[70] M. Hachkar, B. Beden, C. Lamy, Oscillating electrocatalytic systems: Part I. Survey of systems involving the oxidation of organics and detailed electrochemical investigation of formaldehyde oxidation on rhodium electrodes, J. Electroanal. Chem. 287 (1990) 81–98. Available from: https://doi.org/10.1016/0022-0728(90)87161-C.

[71] J.L. Hudson, T.T. Tsotsis, Electrochemical reaction dynamics: a review, Chem. Eng. Sci. 49 (1994) 1493–1572. Available from: https://doi.org/10.1016/0009-2509(94)85063-1.

[72] M.T.M. Koper, Oscillations and complex dynamical bifurcations in electrochemical systems, in: S.A. Rice (Ed.), Advances in Chemical Physics. I. Prigogine, vol. 92, John Wiley & Sons, New York, 1996, pp. 161–298. Available from: https://doi.org/10.1002/9780470141519.

[73] M. Krausa, W. Vielstich, Potential oscillations during methanol oxidation at Pt-electrodes Part 1: experimental conditions, J. Electroanal. Chem. 399 (1995) 7–12. Available from: https://doi.org/10.1016/0022-0728(95)04249-0.

[74] J. Jiang, A. Kucernak, Electrooxidation of small organic molecules on mesoporous precious metal catalysts: II: CO and methanol on platinum–ruthenium alloy, J. Electroanal. Chem. 543 (2003) 187–199. Available from: https://doi.org/10.1016/S0022-0728(03)00046-9.

[75] L. Han, H. Ju, Y. Xu, Ethanol electro-oxidation: cyclic voltammetry, electrochemical impedance spectroscopy and galvanostatic oscillation, Int. J. Hydrogen Energy 37 (2012) 15156–15163. Available from: https://doi.org/10.1016/j.ijhydene.2012.08.034.

[76] M.T.M. Koper, M. Hachkar, B. Beden, Investigation of the oscillatory electro-oxidation of formaldehyde on Pt and Rh electrodes by cyclic voltammetry, impedance spectroscopy and the electrochemical quartz crystal microbalance, J. Chem. Soc. Faraday Trans. 92 (1996) 3975–3982. Available from: https://doi.org/10.1039/FT9969203975.

CHAPTER FIVE

Production of hydrogen by the electrocatalytic oxidation of compounds derived from the biomass (glycerol, glucose)

Contents

As presented in the introduction (see Chapter 1), the use of biomass compounds, such as bio-derived alcohols, polyols, sugars and so on, allowed to increase the efficiency of the electrolysis process for H_2 production, to decrease the energy consumption, and to produce value-added compounds. Biomass represents a huge potential resource as renewable feedstock for the production of compounds for both fine chemicals and energy applications [1]. Indeed, the chemical composition of biomass molecule $C_xH_yO_z$ makes them highly functionalized and therefore electrochemically reactive for the coproduction of hydrogen and reaction compounds. Among many compounds from biomass, glycerol, which can now be considered as a waste from the biofuel industry, and glucose have already been proposed as interesting hydrogen source [2,3].

The reactions of the complete conversion of glycerol (Eq. (5.1)) and glucose (Eq. (5.2)) into CO_2 and H_2, as examples, are as follows:

$$CH_2OH - CHOH - CH_2OH + 3\,H_2O \rightarrow 3\,CO_2 + 7\,H_2 \tag{5.1}$$

$$\begin{aligned} &CH_2OH - CHOH - CHOH - CHOH - CHOH - CHO \\ &\quad + 6\,H_2O \rightarrow 6\,CO_2 + 12\,H_2 \end{aligned} \tag{5.2}$$

Production of Clean Hydrogen by Electrochemical Reforming of Oxygenated Organic Compounds.
DOI: https://doi.org/10.1016/B978-0-12-821500-5.00005-9

Both these reactions show clearly that these molecules are very good hydrogen carriers. However, owing to the complex structure of the molecules, with more than three carbon-bearing alcohol and/or aldehyde groups, the reaction mechanism is also very complex and will depend on the medium (acidic or alkaline) and on the electrode composition/structure. The cathode of an electrolysis cell is generally based on platinum nanoparticles dispersed on a carbon support, independently of the medium (acidic or alkaline). The working conditions at the anode of electrolysis cells in the electro-conversion of molecules from biomass are much softer than those encountered in water electrolysis cells, with electrode potentials lower than 1.0 V versus reversible hydrogen electrode (RHE), leading to cell voltages smaller than 1.0 V. Under such conditions, the durability of materials is not a major concern, and catalysts consisting generally in metal nanoparticles dispersed on a carbon support sustain sufficient chemical and thermal stabilities. However, the choice of the anodic electrocatalysts will be governed by the medium in terms of stability, activity, and selectivity, with the aim of achieving the lowest anode potential possible, that is, to decrease the cell voltage and minimize the electrical energy consumption. Moreover, these catalysts must be allowed to reach either a high oxidation state of glycerol or glucose with the formation of carbon dioxide or to maximize the yield of production of partially oxidized value-added compounds.

5.1 Electrochemical reforming of glycerol

Glycerol, a by-product of the biofuel industry with excess production, presents the advantage of being available for a low price. Its use for hydrogen production and potentially value-added compounds from its electrooxidation would make the biofuel industry more profitable. Moreover, the overall process will have an E-factor, as defined by Sheldon [4], close to 0 because it can be considered as waste from industry using primary renewable resources [5].

5.1.1 Glycerol electrooxidation in acidic media

5.1.1.1 General aspect of glycerol electroreforming in acidic media

The electrooxidation of glycerol for the hydrogen production in a proton exchange membrane fuel cell was first studied by Marshall and

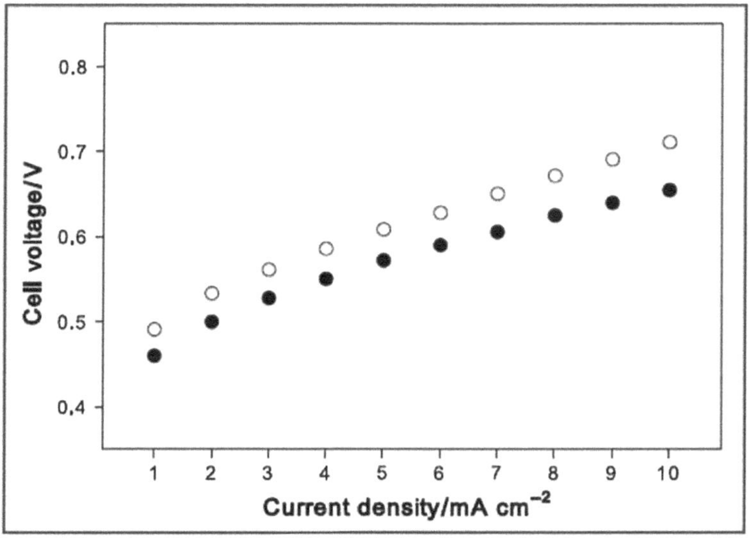

Figure 5.1 Polarization curves of the PEM electrolysis cell at 70 °C with 2 mol L^{-1} (●) and 8.5 mol L^{-1} glycerol (O) circulating through the anodic compartment [6].

Haverkamp [6]. Their system consisted in a classical Pt/C cathode for proton reduction into hydrogen, and a RuO_2-IrO_2/C anode for the oxidation of glycerol, both electrodes having a surface area of 3.14 cm^2, loaded at 3 mg_{metal} cm^{-2}. These authors justified the choice of the anode by the fact that hydrous RuO_2 displayed good activity for methanol electrooxidation, whereas IrO_2 could promote ethanol oxidation. The acidic electrolyte membrane was a Nafion 212 membrane (50 μm thickness). The performance of an electrolysis cell at 70°C fed with a 2 M glycerol aqueous solution reached 10 mA cm^{-2} at a cell voltage of c.0.65 V (Fig. 5.1). Electrical energy consumption of c.1.1 kWh (Nm H_2)$^{-3}$ at an H_2 production rate of 10 A m^{-2} could be achieved. In order to explain the relatively low efficiency of glycerol electro-conversion, they pointed out the high overvoltage for the reaction to start. They also admitted that in their study, the level of glycerol conversion to CO_2 was unknown. Generally, in acidic medium, several compounds are formed from glycerol oxidation. Indeed, Kongjao et al. have detected C1−C6 reaction products with different oxidation levels for the electrochemical reforming of glycerol on Pt anodes [7]. The number of hydrogen molecules recovered per molecule of glycerol oxidized becomes lower than the theoretical one for the complete oxidation of glycerol into CO_2, which decreases the efficiency of the system.

These two catalytic aspects, for example, overvoltage and selectivity, belong to the domain of electrocatalysis. The overvoltage can be decreased by lowering the adsorption strength of reaction intermediates, by activating water molecules at low potentials, and so on, whereas the selectivity toward CO_2 can be improved by favoring the C−C bond breaking. These improvements depend on the catalyst composition and structure.

In acidic media, platinum is unavoidable for the electroactivation of alcohols. Indeed, this material is the most active one for the dissociative adsorption of alcohols at its surface [8,9], allowing further oxidation of the adsorbed species with an acceptable rate. Recently, Valter et al. have shown, using combined experimental and density functional theory (DFT) studies, that gold could have a small activity toward glycerol electrooxidation in the acidic medium [10]. These results contradict well those established previously [11−13]. Anyway, the extent of activity on gold surface in acidic medium was so tiny that gold cannot be seriously considered as a viable catalyst for hydrogen production from glycerol electroreforming.

In contrast, Roquet et al. studied the glycerol electrooxidation on a smooth platinum sheet of 10 cm^2 geometric surface area with a real surface area of 40 cm^2 [14]. They found that in an acidic medium, the dissociative oxidation of glycerol leading to formic acid, oxalic acid, and glycolic acid was the main process at high electrode potentials where the Pt surface is covered by oxides, whereas at lower potentials, very high selectivity toward glyceraldehyde was obtained. These results were confirmed on platinum nanoparticles dispersed on carbon support (Pt/C) by Kwon et al. based on the analysis by high-performance liquid chromatography (HPLC) of reaction products collected during the oxidation of 0.1 M glycerol in a 0.5 M H_2SO_4 electrolyte solution under linear sweep voltammetry [15]. These authors obtained a very high selectivity toward the formation of glyceraldehyde at potentials from 0.4 to 0.9 V versus RHE, with also the formation of glyceric acid, according to Eqs. (5.3) and (5.4):

$$CH_2OH-CHOH-CH_2OH \rightarrow CH_2OH-CHOH-CHO+2H^++2e^- \tag{5.3}$$

$$CH_2OH-CHOH-CH_2OH+H_2O \rightarrow CH_2OH-CHOH-COOH+4H^++4e^- \tag{5.4}$$

For potentials higher than 1.0 V, that is, when the Pt nanoparticles are covered by surface oxides, the selectivity toward glyceraldehyde decreases in favor of the formation of compounds with higher oxidation levels such as first glyceric acid, then formic acid, and glycolic acid. They found an increase in glycolic acid and formic acid selectivities together with a strong decrease in selectivity in glyceraldehyde and glyceric acid, which seems to indicate that both these latter compounds behave as reaction intermediates toward higher oxidized chemical products. The formation of CO_2 was pointed out only for potentials higher than 1.1 V versus RHE using in situ Fourier transform infrared reflectance spectroscopy (FTIRS) measurements.

Gomes et al. performed very interesting studies on glycerol electrooxidation at the platinum surface in acidic media. First, they investigated the electrooxidation of 0.1 M glycerol in 0.1 M H_2SO_4 and 0.1 M $HClO_4$ electrolytes on low Miller-index Pt crystalline surface orientations, Pt (100), Pt(110), and Pt(111), by cyclic voltammetry and in situ FTIRS [16]. They observed the formation of adsorbed CO species at low electrode potentials on all Pt surfaces. However, the reaction was found to be surface sensitive: Pt(100) and Pt(110) surface structures favored the breaking of the C–C bond leading to adsorbed CO species, whereas the Pt (111) surface structure did not favor the C–C bond breaking at potentials as low as 0.05 V. Moreover, these CO species were found to remain attached to Pt(100) and Pt(110) surface up to higher potentials than on Pt (111). Therefore, Pt(111) surface was less poisoned by adsorbed species from glycerol and showed the highest activity at low potentials. At potentials higher than 0.3 V versus RHE for Pt(111) and 0.5 V for Pt(100) and Pt(110), carbonyl-containing compounds and CO_2 were formed. The comparison of the relative intensities of the CO_2 and carbonyl infrared bands revealed that Pt(100) and Pt(110) were more active than Pt(111) for the production of CO_2, resulting from C–C bond breaking, These authors also investigated the effect of glycerol concentration on the alcohol adsorption and electrooxidation on Pt by cyclic voltammetry and in situ FTIR spectroscopy [17]. They first confirmed that CO was formed at low potentials, competing with hydrogen adsorption on Pt sites. However, they also showed that the electrooxidation of glycerol on Pt is strongly influenced by its concentration, which seems to increase the ability of Pt to produce other adsorbates than CO. High glycerol concentrations led to a Pt surface strongly populated by glycerol residues blocking the formation of oxygenated species on the Pt surface and preventing CO oxidation to CO_2 at higher potentials via a Langmuir–Hinshelwood mechanism—see Eqs. (22d) and (22f).

During the positive scan in the potential range of 0.6 V versus RHE, that is, the onset potential of glycerol oxidation, and 1.0 V versus RHE, they proposed that parallel electrocatalytic reaction pathways occurred leading to the formation of CO_2 and carboxylic acids. The formation of CO_2 by CO oxidation was favored by decreasing the concentration of glycerol.

5.1.1.2 Effect of the catalyst composition on glycerol electroreforming in acidic media

Attempts were done to increase the activity and/or the selectivity of electrocatalysts by modifying their surface by adatoms or by alloying platinum with other atoms. Indeed, it is known that the presence of foreign metal atoms at the Pt surface could enhance the activity and change the selectivity of the oxidation reaction of alcohols. Different actions of the foreign atoms were proposed in the literature according to their nature and the structure of the catalysts. The activity is depending on the balance between the inhibiting effect due to blocking of active sites by strongly adsorbed species such as CO and the catalytic effect due to enhanced adsorption of OH_{ads} on sites adjacent to the foreign metals, that is, third body effect [18]. This process is followed by a bifunctional mechanism between the adsorbed species from glycerol and adsorbed OH species [19] so that free active surface sites become available for further adsorption and oxidation of glycerol. The course of the electrooxidation reaction is also affected by the amount and composition of strongly chemisorbed species [19,20]. Finally, the dilution of Pt surface atoms by the presence of foreign surface atoms changes the nature and the adsorption strength of the adsorbed species involved in the reaction mechanism [21]. Petrii and coworkers [19,20] proposed then that the presence of adatoms led to significant effects on the amount and course of electrooxidation of strongly bonded species and affected the composition of the chemisorbed species and therefore the selectivity of the catalyst. For example, the work of Garcia et al. [22]. on the role of Bi adatoms irreversibly adsorbed on platinum single-crystal electrodes toward the oxidation of glycerol illustrated these assumptions. It was shown that the presence of bismuth on the Pt (111) electrode prevented the adsorption of surface blocking species (CO and other species from glycerol dissociative adsorption), leading to improve the catalytic activity. They also showed that the selectivity was

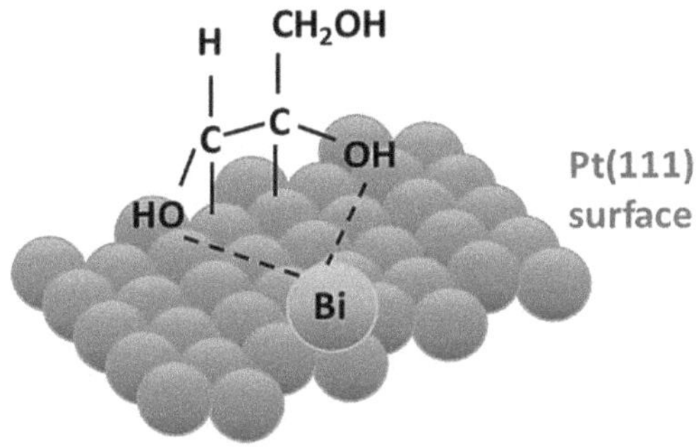

Figure 5.2 Schematic representation of the proposed interaction of the enediol intermediate with Bi on a Pt(111) surface. *Modified after [22].*

orientated toward dihydroxyacetone instead of glyceraldehyde. On a Pt (100)/Bi electrode, the activity was decreased due to strongly adsorbed species, and the selectivity toward glyceraldehyde was not affected. These authors proposed that the enediol intermediate (Fig. 5.2) could only be formed on the Pt(111) surface and that its interaction with a Bi surface adatom was the key parameter for the isomerization reaction between glyceraldehyde and dihydroxyacetone.

Kwon et al [13]. also observed an enhancement of the activity and a 100% selectivity toward dihydroxyacetone on carbon-supported platinum nanoparticles dipped in a bismuth saturated solution at a carefully chosen potential, whereas without bismuth the primary alcohol oxidation toward glyceraldehyde was dominant. Kim et al. compared the activity and the selectivity of a Pt/C catalyst, and alloyed $Pt_{0.7}Ru_{0.3}/C$, $Pt_{0.5}Ru_{0.5}/C$, and $Pt_{0.3}Ru_{0.7}/C$ (atomic ratio) catalysts toward the electrooxidation of 2 M glycerol in 0.5 M H_2SO_4 electrolyte by cyclic voltammetry [23]. The $Pt_{0.5}Ru_{0.5}/C$ displayed the highest activity as well as improved stability for glycerol electrooxidation. The product distribution, after chronopotentiometric measurements of the oxidation of 0.1 M glycerol in 0.5 M H_2SO_4 on Pt/C and $Pt_{0.5}Ru_{0.5}/C$, with 1.0 mg cm^{-2} metal loading, at a reaction temperature of 60 °C and a fixed potential of 1.1 V versus standard hydrogen electrode for 7-hour reaction time, was analyzed by HPLC. The current density recorded after 7 hours of experiment was five times higher on Pt_5Ru_5/C than on Pt/C leading to a glycerol conversion of 31.5% and 20.3%, respectively, and confirming the higher activity of the former catalyst. Although the current density was five times higher, the conversion of glycerol was only c.1.5 times higher with $Pt_{0.5}Ru_{0.5}/C$ than with Pt/C. This discrepancy was due to differences in reaction

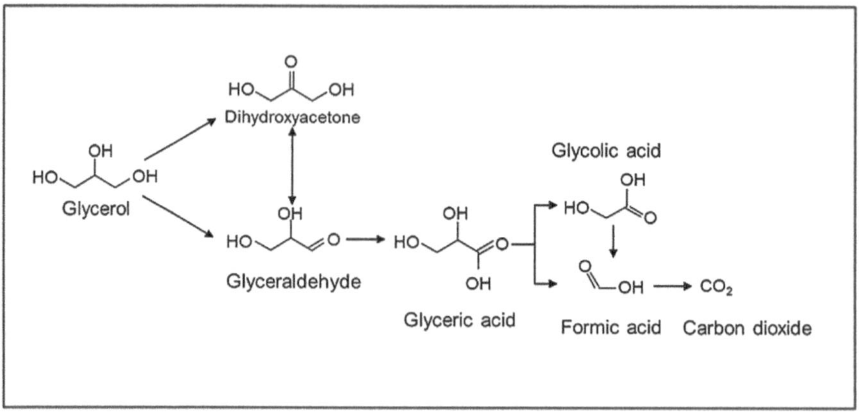

Figure 5.3 Reaction pathways for the electrooxidation of glycerol in acidic media.

product distributions. The authors detected glyceraldehyde, glyceric acid, and glycolic acid with both catalysts with different selectivities, but dihydroxyacetone only with $Pt_{0.5}Ru_{0.5}/C$. The mass balance was found to be c.100% with Pt/C and c.80% with $Pt_{0.5}Ru_{0.5}/C$, indicating that the second catalyst was able to increase the rate of C–C bond breaking to form C1 species, such as formic acid and CO_2.

From these studies, and considering that hydroxypyruvic acid, tartronic acid, and mesoxalic acid were never detected in acidic media, the reaction pathway given in Fig. 5.3 can be proposed for the oxidation of glycerol in acidic media.

It has also been reported that a Pt_2Ni/C catalyst prepared by the polyol method improved the electrocatalytic activity for glycerol electrooxidation and the stability during chronoamperometric measurements [24]. However, although the modification of Pt by foreign metal atoms allowed increasing the activity and affecting the selectivity, the onset potentials for glycerol electrooxidation in acidic medium were only slightly decreased, ranging from c.0.6 V versus RHE for platinum to c.0.4 V versus RHE for modified catalysts, that is, far from the reversible potential of glycerol oxidation into CO_2 (see Table 2.1 in Section 2.1). This is due to the necessity to use platinum, which is very active in an acidic medium for the dissociative adsorption of alcohols and particularly glycerol, but very prone to be poisoned by strongly adsorbed species. These species cannot be removed from the electrode surface except if platinum or a more oxophilic metal is able to activate, at lower potentials, the water molecules necessary to provide extra oxygen atoms to oxidize completely the molecule according to the so-called bifunctional mechanism [25]. However, the onset potential of glycerol oxidation must be lowered to decrease the cell voltage and thus the electrical energy consumption to produce hydrogen using glycerol in a proton exchange membrane electrolysis cell (PEMEC). The second limitation comes from the fact that glycerol electrooxidation in acidic medium leads to a wide distribution of products and that production of CO_2 remains low even with modified catalysts. Table 5.1 gives the number of moles of hydrogen produced for the oxidation of glycerol into different compounds. A lower hydrogen amount is produced when C3, C2, and C1 compounds other than CO_2 are formed. Because low amounts of CO_2 are formed, fewer hydrogen molecules are produced per molecules of oxidized glycerol, and the efficiency of the glycerol PEMEC will also become low.

5.1.2 Glycerol electrooxidation in alkaline media

In order to overcome some limitations encountered in acidic media, the electrooxidation of glycerol in alkaline media can be considered. Indeed, it is well known that the kinetics of alcohol oxidation is faster in alkaline media than in acidic ones [14,26], and that non-noble metal catalysts are more stable. Therefore, in addition to classical Pt-based catalysts, low platinum loadings or even nonplatinum-based catalysts displaying high activity as anode materials in alkaline media could be developed. Hence, the electrooxidation of glycerol has been studied on platinum [27–29], palladium [2,27,28,30,31], gold [32–35], nickel [36–39], and rhodium-based [40] catalysts.

5.1.2.1 General aspect of glycerol electroreforming in alkaline media

Studies of glycerol electrooxidation by linear sweep voltammetry on pure Pt/C, Pd/C, and Au/C [30,32] showed that platinum makes it possible to oxidize glycerol from an onset potential, of c.0.4 V versus RHE, lower than do it palladium and gold at c.0.55 V versus RHE for both catalysts. Thus, two different reaction mechanisms could be involved from the onset potential of glycerol oxidation up to c.0.8 V versus RHE. Fig. 5.4 displays polarization curves of glycerol electrooxidation recorded on pure Pt/C, Pd/C, and Au/C.

On Pt/C and Pd/C, the products detected were mainly glyceraldehyde, glycerate, tartronate, and/or mesoxalate ions, whereas on Au/C hydroxypyruvate was also detected in addition to the other carboxylates [32]. Moreover, the infrared absorption band relative to the formation of

Table 5.1 Number of hydrogen moles formed per mole of glycerol oxidized in an electrolysis cell.

Electrolysis reaction	Mole H_2	Products
$CH_2OH - CHOH - CH_2OH \rightarrow CH_2OH - CHOH - CHO +$	$1\ H_2$	Glyceraldehyde
$CH_2OH - CHOH - CH_2OH \rightarrow CH_2OH - CO - CH_2OH +$	$1\ H_2$	Dihydroxyacetone
$CH_2OH - CHOH - CH_2OH + H_2O \rightarrow CH_2OH - CHOH - COOH +$	$2\ H_2$	Glyceric acid
$CH_2OH - CHOH - CH_2OH + 2\ H_2O \rightarrow CH_2OH - COOH + HCOOH +$	$3\ H_2$	Glycolic acid/Formic acid
$CH_2OH - CHOH - CH_2OH + 2\ H_2O \rightarrow CH_2OH - COOH + CO_2 +$	$4\ H_2$	Glycolic acid/carbon dioxide
$CH_2OH - CHOH - CH_2OH + 3\ H_2O \rightarrow 3\ HCOOH +$	$4\ H_2$	Formic acid
$CH_2OH - CHOH - CH_2OH + 3\ H_2O \rightarrow 2\ HCOOH + CO_2 +$	$5\ H_2$	Formic acid/carbon dioxide
$CH_2OH - CHOH - CH_2OH + 3\ H_2O \rightarrow HCOOH + 2\ CO_2 +$	$6\ H_2$	Formic acid/carbon dioxide
$CH_2OH - CHOH - CH_2OH + 3\ H_2O \rightarrow 3\ CO_2 +$	$7\ H_2$	Carbon dioxide

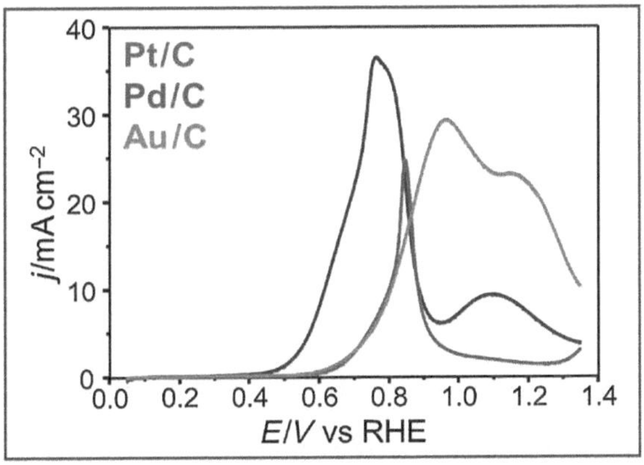

Figure 5.4 Polarization curves of glycerol oxidation recorded on Pt/C, Pd/C, and Au/C catalysts. $v = 10$ mV s^{-1}, N_2-purged, 0.1 mol dm^{-3} glycerol + 1.0 mol dm^{-3} NaOH solution, $T = 293$ K.

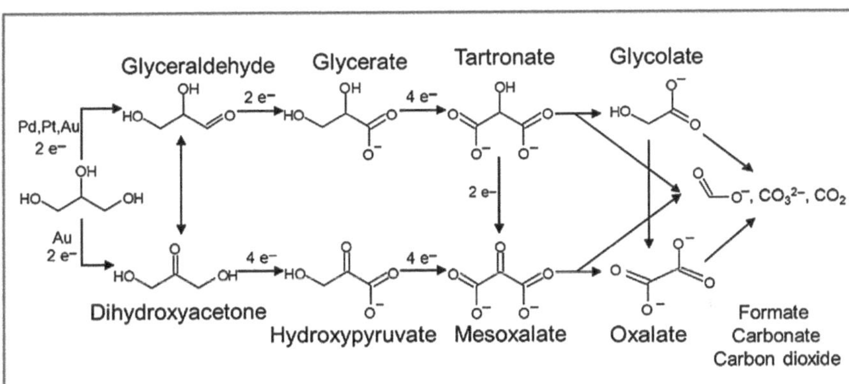

Figure 5.5 Reaction pathway for the electrooxidation of glycerol in alkaline medium.

hydroxyacetone was observed with all catalysts, but with a higher relative intensity on Au/C. These results were confirmed by other studies [2,27,28,30,31]. At higher potentials, the formation of CO_2 occurred indicating the breaking of the C−C bonds and therefore the destruction of the carbon chain. From these observations, a mechanism involving mainly the activation of the primary alcohol was proposed to occur on Pt/C and Pd/C, whereas a mechanism involving the activation of the secondary alcohol group could occur to a large extent on Au/C, according to the scheme given in Fig. 5.5. The fact that dihydroxyacetone is also detected by in situ infrared spectroscopy on Pt and Pd, whereas no hydroxypyruvate is formed at higher potentials, indicates that the equilibrium of an isomerization reaction between dihydroxyacetone and glyceraldehyde is displaced toward the formation of the aldehyde on the Pt/C and Pd/C catalysts comparatively to the Au/C catalyst. In the case of Pt and Pd materials, the oxidation reaction kinetics being much higher than that of the isomerization reaction, the detection of aldehydes by the in situ infrared technique becomes very difficult. The nature of the metal then tends to orient the reaction pathway.

Ni-based catalysts are known to be active for the electrooxidation of alcohols in alkaline media so that this material was used for studying the electrooxidation of glycerol. In this case, glycerol oxidation begins simultaneously with the $Ni^{II}(OH)_2 \rightarrow Ni^{III}OOH$ transition at high potentials. Therefore, the oxidation of glycerol can only occur at potentials higher than 1.0 V versus RHE, that is, in a very oxidative potential range. The main reaction products detected on nickel-based catalysts are then carboxylate ions, which are mostly glycerate. At high electrode potentials formate and carbon dioxide are also produced although in relatively small amounts [37,39]. Glycerate ions seem to be involved in both reactions as products and/or reaction intermediates for the formation of smaller molecules after the C−C bond breaking. The high electrode potentials needed to carry out the oxidation reaction of glycerol will lead to an increase in the electrical energy consumption to produce hydrogen in a glycerol electrolysis cell, which makes nickel-based catalysts not suitable for producing hydrogen by electrolysis at a low operational cost.

5.1.2.2 Effect of the catalyst composition on glycerol electroreforming in alkaline medium

In alkaline media, platinum still displays a better activity at low potentials than the other metals. But, in order to compete with this metal, numerous studies were devoted to the modification of palladium by foreign atoms. For example, because gold also displays an interesting activity in alkaline media, Pd_xAu_{1-x}/C catalysts were developed and evaluated for the electrooxidation of glycerol [32,41]. The enhancement of the catalytic activity, in terms of a decrease in the oxidation onset potential, by c.50 mV, was observed for the Pd−Au alloys comparatively to the pure metals (Fig. 5.6). However, the selectivity of the catalysts was not affected. The main reaction route corresponds first to the formation of glyceraldehyde as for pure Pd. The shift of the onset potential allows better controlling the selectivity toward glycerate

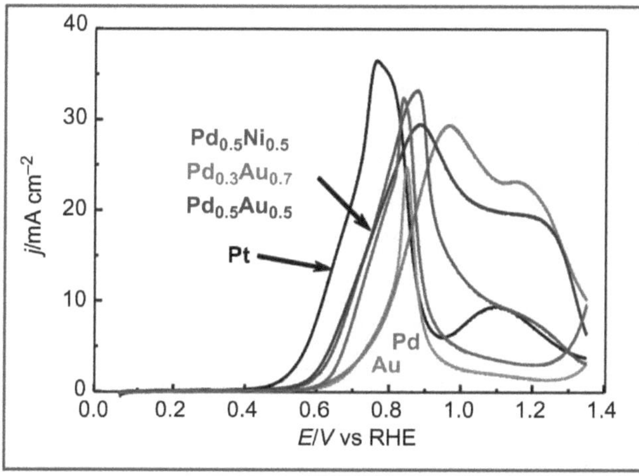

Figure 5.6 Polarization curves of glycerol oxidation recorded on Pt/C, Pd/C, Au/C, PdAu/C, and PdNi/C catalysts ($v = 10\ mV\ s^{-1}$, N_2-purged, $0.1\ mol\ dm^{-3}$ glycerol + $1.0\ mol\ dm^{-3}$ NaOH solution, $T = 293$ K).

formation below 0.8 V, but for higher potentials, about the same product distribution was observed as in the case of pure Pd.

Many other elements from p- and d-groups were used to modify palladium, such as indium [42], tin [43,44], bismuth [45–47], nickel [32,48], silver [48], and so on, have been studied for the electrooxidation of glycerol in alkaline media. For most of these co-catalysts, an increase in activity was observed only from the potential where palladium started to be active and the selectivity was not significantly affected, and only higher current densities were recorded from the onset potential of glycerol electrooxidation. Tin and bismuth as Pd co-catalysts led to very different behaviors with a shift of the onset potential from c.0.6 V versus RHE for pure Pd/C catalyst to c.0.45 V versus RHE for Pd-Sn [43] and 0.4 V versus RHE for Pd-Bi [28]. A catalyst with a Pd/Bi atomic composition of 9:1 leads to the same activity as platinum (Fig. 5.7(A)). However, different selectivities were observed in both materials. In the case of Pd-Bi catalysts, the same mechanism as for platinum was observed with the formation of glyceraldehyde and glycerate at potentials lower than 0.8 V versus RHE and carboxylates, formate, and CO_2 at higher potentials. In the case of Pd-Sn catalysts, the formation of glycerate occurs as soon as the catalyst becomes active for the glycerol electrooxidation and no evidence of the formation of C1 compounds was observed at high potentials. This means

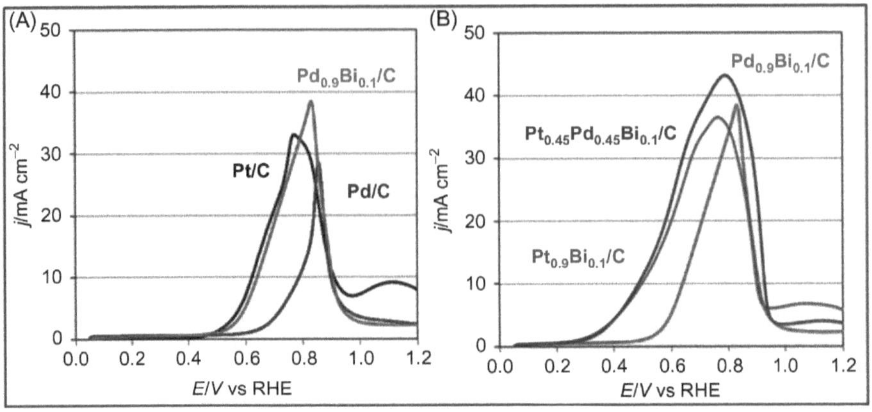

Figure 5.7 Polarization curves of glycerol oxidation recorded on (A) Pt/C, Pd/C, and $Pd_{0.9}Bi_{0.1}/C$ and (B) $Pt_{0.9}Bi_{0.1}/C$, $Pd_{0.9}Bi_{0.1}/C$, and $Pt_{0.45}Pd_{0.45}Bi_{0.1}/C$ ($v = 10$ mV s^{-1}, N_2-purged, 0.1 mol dm^{-3} glycerol + 1.0 mol dm^{-3} NaOH solution, $T = 293$ K).

that Pd-Sn is very selective toward the formation of C3 carboxylates without C−C bond breaking.

The modification of Pt and $Pt_{0.5}Pd_{0.5}$ alloy by bismuth was also investigated. $Pt_{0.9}Bi_{0.1}/C$ and $Pt_{0.45}Pd_{0.45}Bi_{0.1}/C$ catalysts have been prepared by the water-in-oil microemulsion method and compared with Pd/C and Pt/C catalysts with respect to their activity for the electrooxidation of glycerol (Fig. 5.7(B)).

The modification of Pt by Bi led to a decrease in the onset potential of the glycerol oxidation reaction down to 0.2 V versus RHE, which makes this catalyst the best one for this reaction. The replacement of half of the platinum atoms by palladium atoms does not change the activity at low potentials. The reaction mechanism of glycerol oxidation on these catalysts was investigated by in situ infrared spectroscopy (Fig. 5.8).

The infrared spectra in Fig. 5.8(A) show four main infrared absorption peaks. The first one appears as early as 0.05 V versus RHE at c.1225 cm^{-1} and shifts up to c.1265 cm^{-1} from 0.05 to 0.6 V versus RHE. This peak was assigned to the formation of glyceraldehyde, probably in an adsorbed state from 0.05 to 0.6 V, explaining the wavenumber shift, and of other reaction products at higher potentials. The absorption peaks located at c.1310, 1385, and 1570 cm^{-1} were assigned to the formation of glycerate and other carboxylate ions. In the cases of the binary $Pt_{0.9}Bi_{0.1}/C$ and ternary $Pt_{0.45}Pd_{0.45}Bi_{0.1}/C$ catalysts, the shift of the glycerol oxidation onset potential down to 0.2 V, observed in Fig. 5.8(B), is

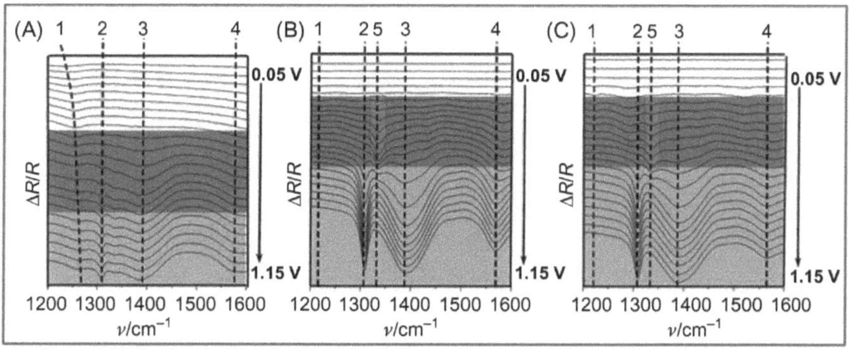

Figure 5.8 Infrared spectra recorded in the 1200–1600 cm^{-1} range during glycerol oxidation from 0.05 to 1.15 V versus RHE on carbon-supported catalysts prepared by the water-in-oil microemulsion method. (A) $Pd_{0.9}Bi_{0.1}/C$, (B) $Pt_{0.9}Bi_{0.1}/C$, and (C) $Pt_{0.45}Pd_{0.45}Bi_{0.1}/C$ catalysts in 0.1 mol dm^{-3} glycerol, 1.0 mol dm^{-3} NaOH solution at 20 °C. Scan rate: 1 mV s^{-1}, resolution 4 cm^{-1}.

accompanied by the formation with high selectivity, at potentials from 0.25 to 0.60 V versus RHE, of dihydroxyacetone, as seen by an absorption band at c.1335 cm^{-1}, and glyceraldehyde with an absorption band at c.1225 cm^{-1}. The activation of water at the surface of these catalysts is likely to be ineffective at very low electrode potentials, which explains the absence of over-oxidation of glycerol into carboxylate. It has been then proposed that bismuth led changing either the electronic state of platinum and/or the glycerol adsorption mode controlling the selectivity toward the formation of ketone or aldehyde species. Another very important effect of modification of platinum by bismuth is that the absorption bands between c.1950 and 2000 cm^{-1} assigned to adsorbed CO species is visible on Pt/C and not on $Pt_{0.9}Bi_{0.1}/C$ (Fig. 5.9). This means that the $Pt_{0.9}Bi_{0.1}/C$ catalyst is less efficient to activate the C–C bond breaking than the Pt/C one and further leads to a higher selectivity in the formation of C3 compounds.

5.1.2.3 Hydrogen production in an alkaline electrolysis cell

From these results, the electrochemical reforming of polyol in alkaline media becomes very questionable considering the objectives of the hydrogen production at potentials below 1.0 V, with a high rate, that is, a high current density, and with the recovery of valuable coproducts from polyol oxidation reaction corresponding to no C–C bond breaking.

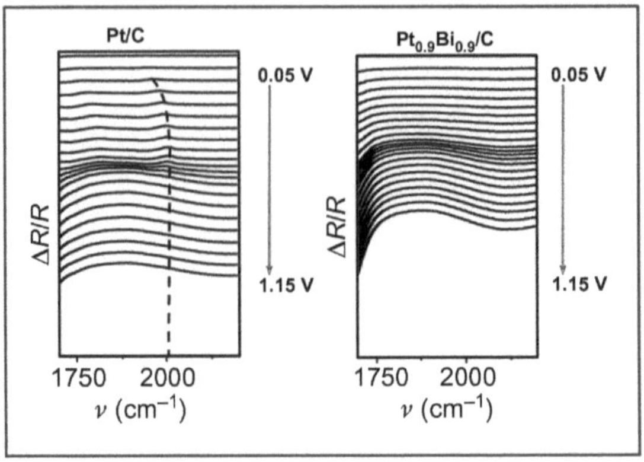

Figure 5.9 Infrared spectra recorded in the 1700–2300 cm^{-1} range during glycerol oxidation from 0.05 to 1.15 V versus RHE on carbon-supported catalysts Pt/C and $Pt_{0.9}Bi_{0.1}/C$ catalysts in 0.1 mol dm^{-3} glycerol, 1.0 mol dm^{-3} NaOH solution at 20 °C. Scan rate: 1 mV s^{-1}, resolution 4 cm^{-1}.

Gonzales-Cobos et al. compared the performances of a glycerol alkaline electrolysis cell fitted with Pt/C, $Pt_{0.9}Bi_{0.1}/C$, and $Pt_{0.3}Pd_{0.6}Bi_{0.1}/C$ anodes and a Pt/C cathode in a 5 cm^2 electrolysis cell [27]. The metal loading of the electrodes was 1.6 mg cm^{-2} and their Nafion loading was c.0.8 mg cm^{-2}. The electrodes, that is, Pt/C cathode and Pt/C, $Pt_{0.9}Bi_{0.1}/C$, or $Pt_{0.3}Pd_{0.6}Bi_{0.1}/C$ anodes, were separated by a simple blotting paper and mechanically pressed in the cell. Electrolysis measurements carried out at 293 K in a 0.5 M NaOH + 2 M glycerol solution, at $U_{cell} = 0.55$ and 0.7 V, show comparatively the current intensity I versus time t, that is, the hydrogen production rates (Fig. 5.10). The order of activity is Pt/C < $Pt_{0.3}Pd_{0.6}Bi_{0.1}/C$ < $Pt_{0.9}Bi_{0.1}/C$, giving stabilized current of c.0.025, 0.035, and 0.05 A, respectively, after 240 minutes at 0.55 V. At 0.7 V, all catalysts display the same activity of c.80 mA after a 60-minute electrolysis. However, the activity of Pt/C decreases drastically for longer times of experiments, likely due to surface poisoning.

Fig. 5.11 shows the chronopotentiometric curves recorded with a $Pt_{0.9}Bi_{0.1}/C$ anode at 0.55 and 0.7 V in a glycerol electrolysis cell fed with a 2 M glycerol + × M NaOH aqueous solutions with $x = 0.5$, 1.0, and 2.0 M. The cell performances are enhanced with the increase in the NaOH concentration, as higher currents, that is, higher hydrogen evolution rates, are obtained. With 2.0 M glycerol in 2.0 M NaOH, hydrogen production rates of c.0.12 Nm^3 H_2 h^{-1} m^{-2} and 0.25 Nm^3 H_2 h^{-1} m^{-2}

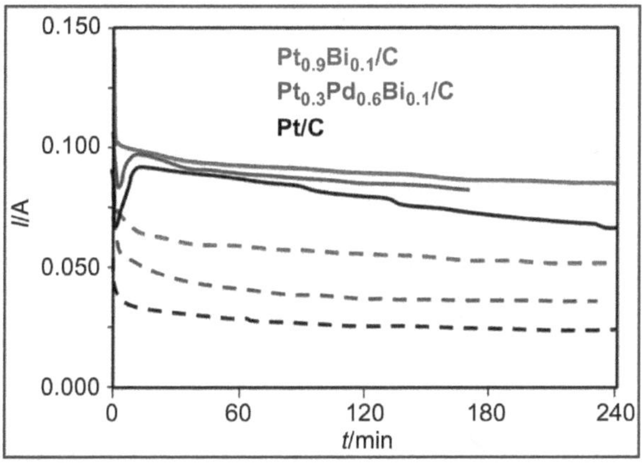

Figure 5.10 Electrolysis cell current versus time at 293 K for an electrolysis cell fitted with a Pt/C cathode and different anodes (black lines: Pt/C, red lines: $Pt_{0.9}Bi_{0.1}/C$ and blue lines: $Pt_{0.3}Pd_{0.6}Bi_{0.1}/C$). 2 M glycerol + 0.5 M NaOH, flow rate = 2 mL min^{-1}; U_{cell} = 0.55 V (dashed lines) and 0.7 V (solid lines). *After [27].*

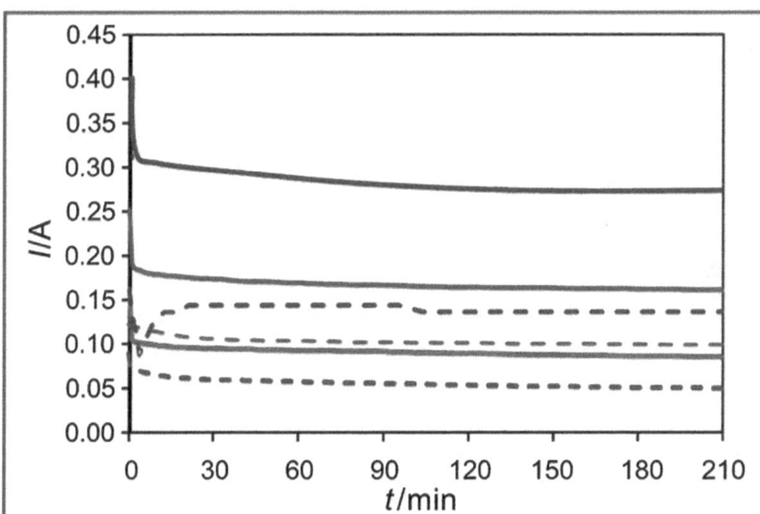

Figure 5.11 Electrolysis cell current versus time at 293 K for an electrolysis cell fitted with a Pt/C cathode and a $Pt_{0.9}Bi_{0.1}/C$ anode and fed with 0.5 M NaOH + 2 M glycerol (red lines), 1.0 M NaOH + 2 M glycerol (blue lines), 2.0 M NaOH + 2 M glycerol (green lines). Flow rate = 2 mL min^{-1}; U_{cell} = 0.55 V (dashed lines); and 0.7 V (solid lines). *After [27].*

Table 5.2 Yields in reaction products determined after 4 h electrolysis experiments at 293K in an electrolysis cell fitted with a Pt/C cathode and a $Pt_{0.9}Bi_{0.1}/C$ anode and fed with 2 M glycerol + 0.5 M NaOH.

Cell voltage	**0.55 V**		**0.7 V**	
Reaction product	**Yield/%**	**Q_{theor}/C**	**Yield/%**	**Q_{theor}/C**
Glyceraldehyde	79.6	748	58.3	606
Glycerate	10.2	210	23.6	493
Hydroxypyruvate	4.1	89	7.2	244
Tartronate	6.1	242	10.9	444
Total	100	1289	100	1787

were achieved with an electrical energy consumption of 1.21 kWh (Nm^3 H_2)$^{-1}$ and 1.54 kWh (Nm^3 H_2)$^{-1}$ at 0.55 and 0.7 V, respectively.

In summary, the production of valuable coproducts such as glyceraldehyde, dihydroxyacetone, glycerate, and/or hydroxypyruvate, besides that of hydrogen, was achieved on Pt-based catalysts in acidic media and on several Pt-based and Pd-based catalysts in alkaline media. For this purpose, the cell voltage was not increased over 0.7 V to avoid a C–C bond breaking observed over 0.8 V on such catalysts. However, this needs relatively high potentials to produce hydrogen with high rate, that is, electrolysis cell voltage greater than 1.0 V leading to high operational cost. Moreover, the product distribution, as determined by chromatographic analysis of the reaction medium after electrolysis running for a long time, showed a very high selectivity into the formation of glyceraldehyde at 0.55 V and its transformation into carboxylate ions at 0.70 V (Table 5.2). The high selectivity toward a value-added compound at a low cell voltage could also decrease the production cost of hydrogen.

Chen et al. prepared Pd nanoparticles supported on a three-dimensional (3D)-nanostructured TiO_2 nanotube arrays (TNTA cloth) and tested this catalyst for glycerol electroreforming in an electrolysis cell equipped with an anion exchange membrane (Tokuyama A201) and a Pt/C on carbon cloth cathode with a loading of 0.3 mg_{Pt} cm^{-2} [49]. The performances of the electrolysis cell fed with 2 M glycerol in a 2 M NaOH aqueous solution at 80 °C reached 1 A cm^{-2} at 0.76 V, which corresponds to an energy consumption of c.1.67 kWh (Nm^3 H_2)$^{-1}$ or c.20.5 kWh (kg H_2)$^{-1}$.

The values of energy consumption obtained by Gonzales-Cobos et al. [27] and Chen et al. have to be compared to those of the state-of-the-art water PEM and alkaline electrolyzers, which range [32,49] from c.4 to

6 kWh (Nm^3 H_2)$^{-1}$, that is, 50–70 kWh (kg H_2)$^{-1}$ [50], as well as to the US Department of Energy's target of 45 kWh (kg H_2)$^{-1}$, to understand the real interest of using such biosourced compounds for both the production of hydrogen and value-added chemicals.

5.2 Glucose electrochemical reforming

Cellulose, which is considered as an inedible agriculture waste, is of great interest for the development of a more sustainable chemistry. Indeed, the use of biomass as a reservoir of renewable carbon appears as a suitable and sustainable alternative to depleting fossil carbon reserves [51]. Moreover, cellulose is the most abundant biopolymer available on earth with a yearly worldwide production estimated to be about 8×10^{10} tons [52]. Cellulose can be depolymerized into monosaccharides, for example, glucose, using biotechnological processes or nonthermal activation treatments such as cold plasma [52,53], ball milling [52], ultrasounds [54], and so on. Therefore, glucose becomes an important and cheap raw material resource to produce clean biosource hydrogen.

5.2.1 Electrochemical reforming of glucose in acidic media

The electrocatalytic oxidation of glucose has been mainly studied on noble metals in neutral media simulating physiological conditions for using glucose/oxygen fuel cells as power sources for cardiac pace makers or to develop glucose sensors. More recently, alkaline media were also investigated to produce added-value compounds. Relatively few studies deal with acidic media [55,56]. The most studied electrocatalytic metals are platinum and gold, because both metals display a catalytic activity for these reactions. Platinum shows good activity toward glucose oxidation in acidic medium, whereas gold exhibits a higher oxidation rate in alkaline medium [57].

Kokoh et al. studied the electrooxidation of glucose in perchloric acid and phosphate buffer media at Au and Pt smooth electrodes [58]. They confirmed that higher activity was obtained at low potentials with Pt than with Au in acidic medium, whereas in neutral medium gold displayed higher activity. The Pt electrode was modified by adatoms, such as Pb, Tl, Bi, and the reaction products in solution as well as in the gas phase after electrolysis measurements were analyzed by HPLC and GC chromatographic techniques. In all cases, gluconic acid was the main reaction product, but other carboxylic acids, for example, glucaric, glucuronic, tartric, glyoxylic, oxalic, glycolic, and formic acids, were detected with concentration distributions depending on the adatom type and surface concentration. These results indicated clearly that platinum was able to break the C—C bond, even in a low extent, producing compounds with higher oxidation levels.

In order to favor oxidation reactions as complete as possible, the transformation of glucose into CO_2 being the target and the issue of the process, it is necessary to bring extra oxygen atoms to the reactive molecule. Therefore, the catalyst used for this reaction should also be able to activate water molecules at low potentials. This poly-functionality of the catalyst will also allow recovering the hydrogen contained in water molecules at potentials far lower than that of the water-splitting reaction.

Tin (Sn) is known to be a very oxophilic metal and a good co-catalyst for the electrooxidation of ethanol on a platinum surface. The comparison of the behavior of a Pt/C and a $Pt_{0.9}Sn_{0.1}/C$ will serve to illustrate the poly-functionality role of the catalyst. Fig. 5.12 shows the polarization curves recorded at $T = 293$ K with a scan rate of 5 mV second^{-1} for the

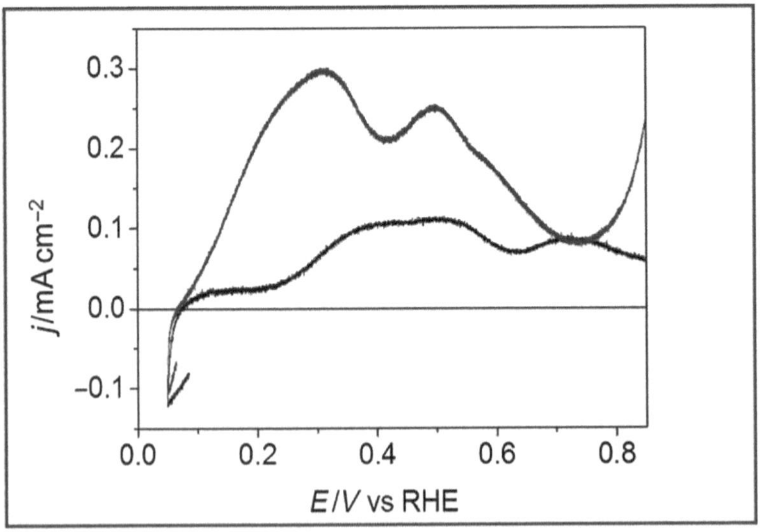

Figure 5.12 Polarization curves of 0.1 M glucose oxidation recorded on Pt/C (black line) and $Pt_{0.9}Sn_{0.1}/C$ (blue line) in 0.1 M H_2SO_4 (scan rate = 5 mV s^{-1}, N_2-purged electrolyte, $T = 293$ K).

electrooxidation of 0.1 M glucose in 0.1 M H_2SO_4 on a Pt/C (black line) and a $Pt_{0.9}Sn_{0.1}/C$ (blue line) prepared by the microemulsion method. The oxidation of glucose starts at c.0.2 V versus RHE on Pt/C catalysts, whereas it starts at c.0.07 V versus RHE on a $Pt_{0.9}Sn_{0.1}/C$ catalyst, that is, a shift of 130 mV toward lower potentials. Moreover, $Pt_{0.9}Sn_{0.1}/C$ leads to higher current density, that is, higher activity, from 0.07 to 0.7 V versus RHE, that is, over the potential range of interest for electrolysis application. This clearly points out the beneficial role of the modification of the Pt surface by tin.

In situ *FTIR spectroscopy* measurements were carried out in order to shed light on the processes occurring at the electrode surface. Fig. 5.13(A) shows the spectra recorded on Pt/C. The spectrum of gluconic acid in the liquid thin film gives two main absorption band at 1637 and 1735 cm^{-1} [59]. The bands located at c.1605 and c.1725 cm^{-1} in the spectra of Fig. 5.13(A) appear simultaneously from c.0.15 V versus RHE and could indicate the formation of carboxylic acids over the whole potential range where the catalysts are active. The second peak shifts from 1715 to 1725 cm^{-1} with a potential increase from c.0.2 to c.1.2 V versus RHE, indicating that the species related to this absorption band is likely to be adsorbed at the Pt surface. In contrast, the peak at c.1637 cm^{-1} seems not to shift with the potential. However, it is also possible that this absorption band is partially hindered by the band due to interfacial water located at

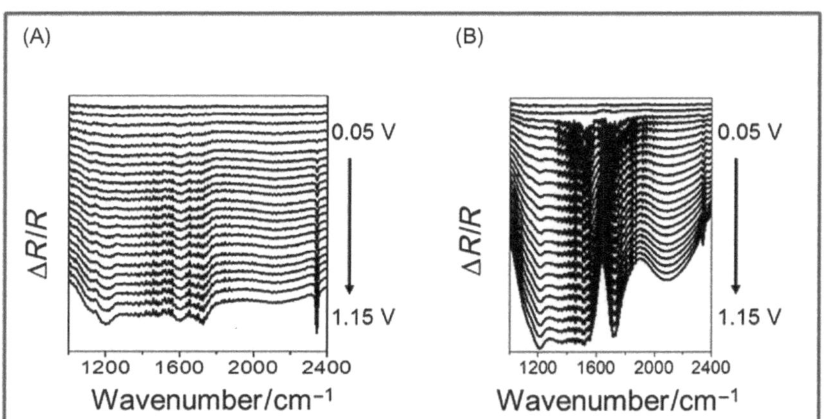

Figure 5.13 Infrared spectra recorded in the 1100–2400 cm^{-1} range during glucose oxidation between 0.05 and 1.15 V versus RHE on carbon-supported catalysts prepared by the water-in-oil microemulsion method: (A) Pt/C and (B) $Pt_{0.9}Sn_{0.1}/C$ (0.1 M glucose, 1.0 M H_2SO_4, 20 °C, scan rate = 1 mV s^{-1}, resolution 4 cm^{-1}).

c.1640 cm^{-1}. Therefore, it is difficult to discriminate the attribution of these bands either to the formation of lactone structures or to carboxylic acids. This peak appears at the same potential as the peak at c.1605 cm^{-1}. However, according to the chromatographic analysis by Kokoh et al. [58], it could be assumed that these bands are mainly due to the formation of gluconic acid. The band at c.2343 cm^{-1} is typical of the formation of interfacial CO_2. This band appears as soon as 0.3 V versus RHE and its intensity increases with the potential, indicating that platinum becomes more and more able to break the C—C bond when increasing the electrode potential.

Fig. 5.13(B) shows the spectra recorded on a $Pt_{0.9}Sn_{0.1}/C$ catalyst. The band due to interfacial water located at c.1640 cm^{-1} is again more pronounced than in the case of the Pt/C catalyst. This is certainly related to the oxophilic character of tin, which reacts with water from very low potentials. The main differences compared with the spectra recorded on Pt/C are that the peak at c.1725 cm^{-1} is much more intense, and the one at c.2343 cm^{-1} is much less intense on $Pt_{0.9}Sn_{0.1}/C$. This means that the presence of tin is detrimental for breaking the C—C bond and favors the production of carboxylic acid, mainly gluconic acid according to the following equation:

$$C_6H_{12}O_6 + H_2O \rightarrow C_6H_{12}O_7 + 2\,H^+ + 2e^- \tag{5.5}$$

5.2.2 Glucose electrochemical reforming in alkaline media

Because alkaline media are less corrosive than the acidic ones, they open wider the range of metals available for the electrooxidation of glucose. Therefore, in addition to platinum [60], which serves as a reference, palladium [61], silver [62], copper [63], nickel [64], and so on have also been used for the electrooxidation of glucose [57]. In alkaline media, gold becomes active for the electrooxidation reaction of organic compounds and has particularly shown very good activity for glucose oxidation [65,66]. However, the highest activities at low electrode potentials were obtained with platinum group metals (PGMs) such as Pt, Pd, and Au.

Fig. 5.14 compares the polarization curves recorded on Pt/C, Pd/C, and Au/C catalysts prepared by the water-in-oil microemulsion method. Pt/C and Au/C lead to the same onset potential for glucose electrooxidation, that is, c.0.2 V versus RHE, whereas Pd/C leads to a higher onset potential of c.0.3 V versus RHE. However, higher activity, that is, higher

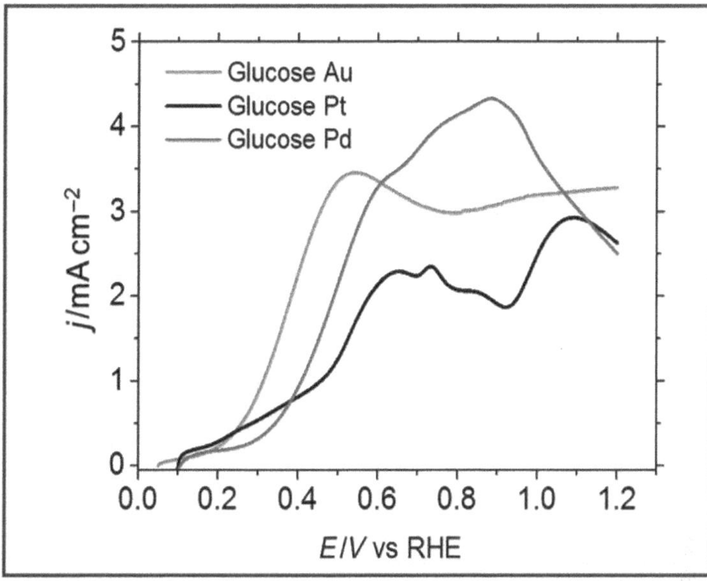

Figure 5.14 Polarization curves of 0.1 M glucose oxidation recorded on a Pt/C (black line), Pd/C (gray line), and Au/C (orange line) in 0.1 M NaOH ($v = 5\ mV\ s^{-1}$, N_2-purged electrolyte, $T = 293$ K).

current densities, is obtained with Au/C than that with Pt/C over the whole potential range from 0.25 to 1.2 V versus RHE. Pd/C displays higher activity than Pt/C from 0.4 V but remains less active than Au/C between 0.2 and 0.6 V versus RHE. The difference in activity could be due to the nature of the species adsorbed on the catalytic surface at the beginning of the oxidation process. According to Beden et al., the formation of strongly adsorbed CO species on platinum tends to deactivate the active surface [60]. The same explanation could be given for palladium. Fig. 5.15(A) represents the infrared spectra recorded on a Pd/C electrode as a function of the electrode potential. A strong absorption band located between 1870 and 1920 cm^{-1} is visible as soon as 0.05 V. This absorption band is typical of CO adsorbed species (CO_{ads}) on a Pd surface [67], which results from the dissociative chemisorption of the glucose molecule. Conversely, in Fig. 5.15(B), which represents the infrared spectra as a function of the electrode potential recorded on an Au/C electrode, no CO adsorption band is observed.

According to Tominaga et al., the deactivation of the gold surface is only due to the formation of close-packed vertically adsorbed gluconic

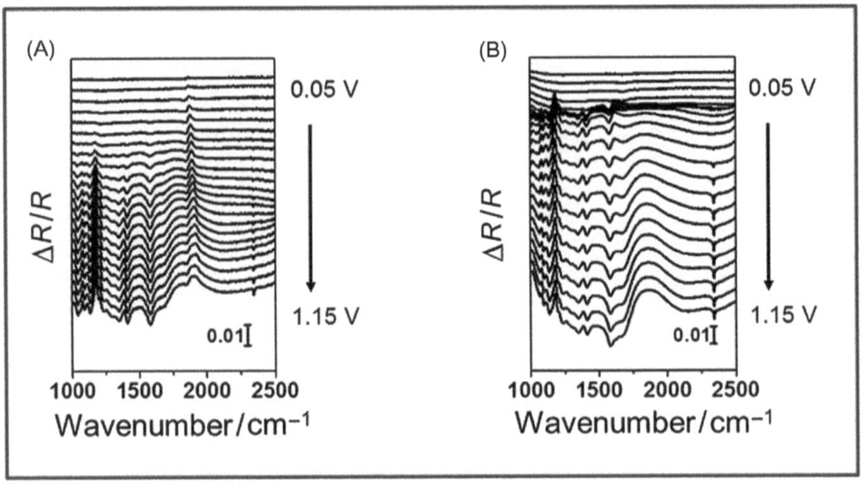

Figure 5.15 Infrared spectra recorded in the 1000–2500 cm^{-1} range during glucose oxidation from 0.05 to 1.15 V versus RHE on carbon-supported catalysts prepared by the water-in-oil microemulsion method: (A) Pd/C and (B) Au/C (0.1 M glucose, 0.1 M NaOH, 20 °C, scan rate = 1 mV s^{-1}, resolution 4 cm^{-1}).

acid molecules [68]. Both catalysts lead to the same infrared absorption peak in the 1000–1750 cm^{-1} wavenumber range, which are typical of the formation of gluconate [60]. Also, CO_2 is formed on both catalysts for potentials higher than 0.65 V versus RHE. But gold displays, in agreement with cyclic voltammetry measurements, higher activity at low potentials, since absorption bands are visible on the spectra for potentials lower than those in the case of Pd/C.

It has been proposed that bimetallic catalysts, particularly Pd–Au materials, could have a better activity for the electrooxidation reaction of glucose [69]. It has indeed been shown by Yan et al. that interactions between Pd and Au in nonalloyed catalysts led to an increase of the tolerance to electrode poisoning [70]. Fig. 5.16 compares the polarization curves recorded for the electrooxidation of 0.1 M glycerol in 0.1 M NaOH aqueous solution on different catalysts, namely Pd/C, Au/C, and alloyed $Pd_{0.3}Au_{0.7}/C$ [71]. The synergetic effect between Pd and Au is clearly demonstrated as the onset potential of glycerol electrooxidation is shifted toward lower values, and higher current densities are recorded in the whole studied potential range.

The $Pd_{0.3}Au_{0.7}/C$ catalyst was used to prepare an anode for a 25 cm^2 electrolysis cell. The electrooxidation of 30 mL of 0.5 M glucose in 0.1 M

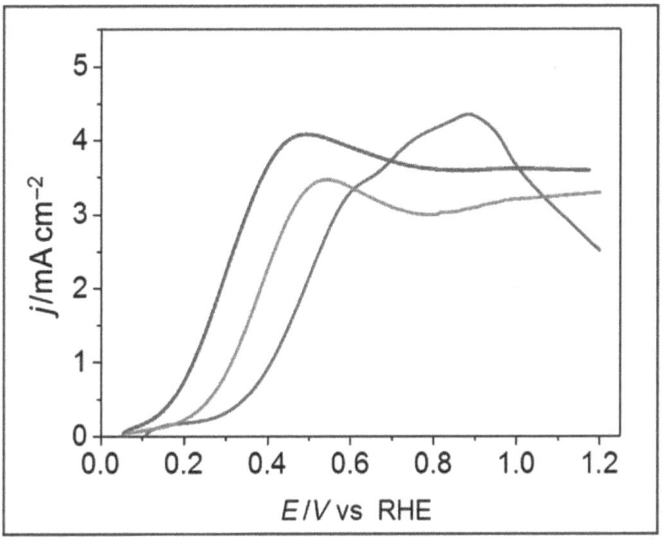

Figure 5.16 Polarization curves of 0.1 M glucose oxidation recorded on a Pd/C (gray line), Au/C (orange line) and $Pd_{0.3}Au_{0.7}/C$ (red line) in 0.1 M NaOH (scan rate = 5 mV s^{-1}, N_2-purged electrolyte, T = 293 K).

NaOH aqueous solution was performed at 293 K for 6 hours at two-cell voltages, for example, 0.4 and 0.8 V, in the electrolysis cell fitted with a $Pd_{0.3}Au_{0.7}/C$ anode (0.5 mg_{metal} cm^{-2}) and a Pt/C cathode (0.5 mg_{Pt} cm^{-2}). Fig. 5.17 shows the current intensity versus time curves, $I(t)$, recorded for the electrolysis experiments.

The initial activity was c.55 mA at 0.4 V and c.65 mA at 0.8 V, which corresponds to initial rates of hydrogen production per square meter of electrode of 0.41 mole H_2 h^{-1} m^{-2}, that is, 0.0094 Nm^3 H_2 h^{-1} m^{-2}, and 0.52 mole H_2 h^{-1} m^{-2}, that is, 0.012 Nm^3 H_2 h^{-1} m^{-2}, respectively. However, the current intensity decreases monotonously with time down to c.20 mA after 6 hours at 0.8 V and c.15 mA after 6 hours at 0.4 V. This decrease in performance with time can be due either to the consumption of glucose with time, which decreases its concentration in the medium or to the deactivation of the catalysts. The total quantity of electricity involved in the 6-hour experiments was 588 and 719 C at 0.4 and 0.8 V, respectively. This quantity of electricity corresponds to the consumption of 0.0030 and 0.0037 mole of glucose, respectively, assuming that two Faradays, corresponding to the production of gluconate, are exchanged per oxidized glucose mole, or less if more than two Faradays are exchanged, that is, less than 0.25% glucose is consumed for 6 hours.

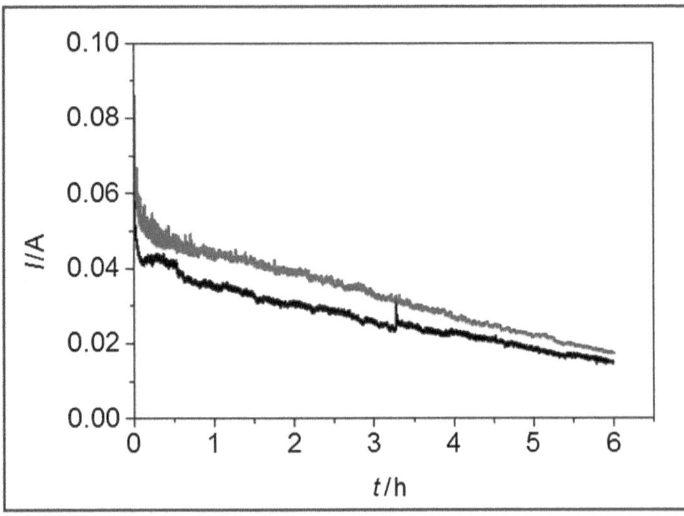

Figure 5.17 Electrolysis current versus time at 293 K in an electrolysis cell fitted with a Pt/C cathode and a $Pd_{0.3}Au_{0.7}/C$ anode. 0.5 M glucose + 0.1 M NaOH, flow rate = 20 mL min^{-1}; U_{cell} = 0.4 V (black line) and 0.8 V (red line).

This assumption was confirmed by HPLC analysis of the reaction products, which showed that the main reaction product was gluconate with a yield of 80% after 1-hour electrolysis and 60% after 6 hours. Therefore, the deactivation of the catalysts seems to be the main reason for the decrease in performances, either due to surface poisoning by adsorbed species from glucose or due to a change in surface composition of the catalyst.

These results show that, although glucose can be oxidized at very low potentials, as low as 0.1 V versus RHE in the case of a $Pd_{0.3}Au_{0.7}/C$ catalyst, the reaction rate of its oxidation and the number of electrons involved per glucose molecule oxidized remain too low to reach high electrolysis current in an electrolysis cell. A very important work on the development of very active electrocatalysts must still be done to allow this green technology to be implemented at a large scale for pure hydrogen production.

References

[1] M.J. Climent, A. Corma, S. Iborra, Converting carbohydrates to bulk chemicals and fine chemicals over heterogeneous catalysts, Green Chem. 13 (2011) 520–540. Available from: https://doi.org/10.1039/c0gc00639d.

[2] V. Bambagioni, M. Bevilacqua, C. Bianchini, J. Filippi, A. Lavacchi, A. Marchionni, et al., Self-sustainable production of hydrogen, chemicals and energy from renewable alcohols by electrocatalysis, ChemSusChem 3 (2010) 851–855. Available from: https://doi.org/10.1002/cssc.201000103.

[3] P.A. Selembo, J.M. Perez, W.A. Lloyd, B.E. Logan, High hydrogen production from glycerol or glucose by electrohydrogenesis using microbial electrolysis cells, Int. J. Hydrogen Energy 34 (2009) 5373–5381. Available from: https://doi.org/10.1016/j.ijhydene.2009.05.002.

[4] R.A. Sheldon, The E factor 25 years on: the rise of green chemistry and sustainability, Green Chem. 19 (2017) 18–43. Available from: https://doi.org/10.1039/c6gc02157c.

[5] M. Aarthy, P. Saravanan, M.K. Gowthaman, C. Rose, N.R. Kamini, Enzymatic transesterification for production of biodiesel using yeast lipases: an overview, Chem. Eng. Res. Des. 92 (2014) 1591–1601. Available from: https://doi.org/10.1016/j.cherd.2014.04.008.

[6] A.T. Marshall, R.G. Haverkamp, Production of hydrogen by the electrochemical reforming of glycerol–water solutions in a PEM electrolysis cell, Int. J. Hydrogen Energy 33 (2008) 4649–4654. Available from: https://doi.org/10.1016/j.ijhydene.2008.05.029.

[7] S. Kongjao, S. Damronglerd, M. Hunsom, Electrochemical reforming of an acidic aqueous glycerol solution on Pt electrodes, J. Appl. Electrochem. 41 (2011) 215–222. Available from: https://doi.org/10.1007/s10800-010-0226-3.

[8] M.P. Hogarth, T.R. Ralph, Catalysis for low temperature fuel cells, part III: challenges for the direct methanol fuel cell, Met. Rev. 46 (2002) 146–164.

[9] J.M. Léger, S. Rousseau, C. Coutanceau, F. Hahn, C. Lamy, How bimetallic electrocatalysts does work for reactions involved in fuel cells? Example of ethanol oxidation and comparison to methanol, Electrochim. Acta 50 (2005) 5118–5125. Available from: https://doi.org/10.1016/j.electacta.2005.01.051.

[10] M. Valter, M. Busch, B. Wickman, H. Grönbeck, J. Baltrusaitis, A. Hellman, Electrooxidation of glycerol on gold in acidic medium: a combined experimental and DFT study, J. Phys. Chem. C 122 (2018) 10489–10494. Available from: https://doi.org/10.1021/acs.jpcc.8b02685.

[11] B. Beden, I. Çetin, A. Kahyaoglu, D. Takky, C. Lamy, Electrocatalytic oxidation of saturated oxygenated compounds on gold electrodes, J. Catal. 104 (1987) 37–46. Available from: https://doi.org/10.1016/0021-9517(87)90334-4.

[12] A. Kahyaoglu, B. Beden, C. Lamy, Oxydation Electrocatalytique Du Glycerol Sur Electrodes d'Or et de Platine En Milieu Aqueux, Electrochim. Acta 29 (1984) 1489–1492. Available from: https://doi.org/10.1016/0013-4686(84)87033-4.

[13] Y. Kwon, K.J.P. Schouten, T.M. Koper, Mechanism of the catalytic oxidation of glycerol on polycrystalline gold and platinum electrodes, Chem Cat Chem 3 (2011) 1176–1185. Available from: https://doi.org/10.1002/cctc.201100023.

[14] L. Roquet, E.M. Belgsir, J.M. Léger, C. Lamy, Kinetics and mechanisms of the electrocatalytic oxidation of glycerol as investigated by chromatographic analysis of the reaction products: potential and pH effects, Electrochim. Acta 39 (1994) 2387–2394. Available from: https://doi.org/10.1016/0013-4686(94)E0190-Y.

[15] Y. Kwon, Y. Birdja, I. Spanos, P. Rodriguez, M.T.M. Koper, Highly selective electro-oxidation of glycerol to dihydroxyacetone on platinum in the presence of bismuth, ACS Catal. 2 (2012) 759–764. Available from: https://doi.org/10.1021/cs200599g.

[16] J.F. Gomes, F.B. Castelo de Paula, L.H.S. Gasparotto, G. Tremiliosi-Filho, The influence of the Pt crystalline surface orientation on the glycerol electro-oxidation in acidic media, Electrochim. Acta 76 (2012) 88–93. Available from: https://doi.org/10.1016/j.electacta.2012.04.1.

[17] J.F. Gomes, C.A. Martins, M.J. Giz, G. Tremiliosi-Filho, G.A. Camara, Insights into the adsorption and electro-oxidation of glycerol: self-inhibition and concentration effects, J. Catal. 301 (2013) 154–161. Available from: https://doi.org/10.1016/j.jcat.2013.02.00.

[18] T.J. Schmidt, R.J. Behm, B.N. Grgur, N.M. Markovic, P.N. Ross Jr., Formic acid oxidation on pure and bi-modified Pt(111): temperature effects, Langmuir 16 (2000) 8159–8166. Available from: https://doi.org/10.1021/la000339z.

[19] B.I. Podlovchenko, O.A. Petry, A.N. Frumkin, H. Lal, The behavior of a platinized-platinum electrode in solutions of alcohols containing more than one carbon atom, aldehydes and formic acid, J. Electroanal. Chem. 11 (1966) 12–25. Available from: https://doi.org/10.1016/0022-0728(66)80053-0.

[20] N.W. Smirnova, O.A. Petrii, A. Grzejdziak, Effect of ad-atoms on the electro-oxidation of ethylene glycol and oxalic acid on platinized platinum, J. Electroanal. Chem. 251 (1988) 73–87. Available from: https://doi.org/10.1016/0022-0728(88)80386-3.

[21] R.R. Adzic, Electrocatalysis on surfaces modified by foreign metal adatoms, in: H. Gerisher, C.W. Tobias (Eds.), Advances in Electrochemistry and Electrochemical Engineering., vol 13, Wiley- Interscience, New York, 1984, pp. 159–260. Available from: http://dx.doi.org/10.1002/bbpc.19860900526.

[22] A.C. Garcia, Y.Y. Birdja, G. Tremiliosi-Filho, M.T.M. Koper, Glycerol electro-oxidation on bismuth-modified platinum single crystals, J. Catal. 346 (2017) 117–124. Available from: https://doi.org/10.1016/j.jcat.2016.12.013.

[23] Y. Kim, H.W. Kim, S. Lee, J. Han, D. Lee, J.R. Kim, et al., The role of ruthenium on carbon-supported PtRu catalysts for electrocatalytic glycerol oxidation under acidic conditions, Chem Cat Chem 9 (2017) 1683–1690. Available from: https://doi.org/10.1002/cctc.201601325.

[24] S. Lee, H.J. Kim, S.M. Choi, M.H. Seo, W.B. Kim, The promotional effect of Ni on bimetallic PtNi/C catalysts for glycerol electrooxidation, Appl. Catal. A: Gen. 429–430 (2012) 39–47. Available from: https://doi.org/10.1016/j.apcata.2012.04.002.

[25] M. Watanabe, S. Motoo, Electrocatalysis by ad-atoms. 2. Enhancement of oxidation of methanol on platinum by ruthenium ad-atoms, J. Electroanal. Chem. 60 (1975) 267–273. Available from: https://doi.org/10.1016/S0022-0728(75)80261-0.

[26] Y. Wang, L. Li, L. Hu, L. Zhuang, J. Lu, B. Xu, A feasibility analysis for alkaline membrane direct methanol fuel cell: thermodynamic disadvantages versus kinetic advantages, Electrochem. Commun. 5 (2003) 662–666. Available from: https://doi.org/10.1016/S1388-2481(03)00148-6.

[27] J. Gonzalez-Cobos, S. Baranton, C. Coutanceau, Development of bismuth-modified PtPd nanocatalysts for the electrochemical reforming of polyols into hydrogen and value-added chemicals, ChemElectroChem 3 (2016) 1694–1704. Available from: https://doi.org/10.1002/celc.201600147.

[28] M. Simões, S. Baranton, C. Coutanceau, Enhancement of catalytic properties for glycerol electrooxidation on Pt and Pd nanoparticles induced by Bi surface modification, Appl. Catal. B: Environ. 110 (2011) 40–49. Available from: https://doi.org/10.1016/j.apcatb.2011.08.020.

[29] R.G. Da Silva, S.A. Neto, K.B. Kokoh, A.R. De Andrade, Electroconversion of glycerol in alkaline medium: from generation of energy to formation of value-added products, J. Power Sources 351 (2017) 174–182. Available from: https://doi.org/10.1016/j.jpowsour.2017.03.101.

[30] M. Simões, S. Baranton, C. Coutanceau, Electrochemical valorisation of glycerol, ChemSusChem 5 (2012) 2106–2124. Available from: https://doi.org/10.1002/cssc.201200335.

[31] W. Wang, Y. Kang, Y. Yang, Y. Liu, D. Chai, Z. Lei, PdSn alloy supported on phenanthroline-functionalized carbon as highly active electrocatalysts for glycerol oxidation, Int. J. Hydrogen Energy 41 (2016) 1272–1280. Available from: https://doi.org/10.1016/j.ijhydene.2015.11.017.

[32] M. Simões, S. Baranton, C. Coutanceau, Electro-oxidation of glycerol at Pd based nano-catalysts for an application in alkaline fuel cells for chemicals and energy cogeneration, Appl. Catal. B: Environ. 93 (2010) 354−362. Available from: https://doi.org/10.1016/j.apcatb.2009.10.008.
[33] D.Z. Jeffery, G.A. Camara, The formation of carbon dioxide during glycerol electrooxidation in alkaline media: first spectroscopic evidences, Electrochem. Commun. 12 (2010) 1129−1132. Available from: https://doi.org/10.1016/j.elecom.2010.06.001.
[34] J.F. Gomes, A.C. Garcia, L.H.S. Gasparotto, N.E. De Souza, E.B. Ferreira, C. Pires, et al., Influence of silver on the glycerol electro-oxidation over AuAg/C catalysts in alkaline medium: a cyclic voltammetry and in situ FTIR spectroscopy study, Electrochim. Acta 144 (2014) 361−368. Available from: https://doi.org/10.1016/j.electacta.2014.08.035.
[35] C.A. Ottoni, S.G. Da Silva, R.F.B. De Souza, A.O. Neto, Glycerol oxidation reaction using PdAu/C electrocatalysts, Ionics 22 (2016) 1167−1175. Available from: https://doi.org/10.1007/s11581-015-1631-8.
[36] Q. Lin, Y. Wei, W. Liu, Y. Yu, J. Hu, Electrocatalytic oxidation of ethylene glycol and glycerol on nickel ion implanted-modified indium tin oxide electrode, Int. J. Hydrogen Energy 42 (2017) 1403−1411. Available from: https://doi.org/10.1016/j.ijhydene.2016.10.011.
[37] V.L. Oliveira, C. Morais, K. Servat, T.W. Napporn, G. Tremiliosi-Filho, K.B. Kokoh, Glycerol oxidation on nickel based nanocatalysts in alkaline medium − identification of the reaction products, J. Electroanal. Chem. 703 (2013) 56−62. Available from: https://doi.org/10.1016/j.jelechem.2013.05.021.
[38] V.L. Oliveira, C. Morais, K. Servat, T.W. Napporn, G. Tremiliosi-Filho, K.B. Kokoh, Studies of the reaction products resulted from glycerol electrooxidation on Ni-based materials in alkaline medium, Electrochim. Acta 117 (2014) 255−262. Available from: https://doi.org/10.1016/j.electacta.2013.11.127.
[39] M.S.E. Houache, E. Cossar, S. Ntais, E.A. Baranova, Electrochemical modification of nickel surfaces for efficient glycerol electrooxidation, J. Power Sources 375 (2018) 310−319. Available from: https://doi.org/10.1016/j.jpowsour.2017.08.089.
[40] M.V. Pagliaro, M. Bellini, M. Bevilacqua, J. Filippi, M.G. Folliero, A. Marchionni, et al., Carbon supported Rh nanoparticles for the production of hydrogen and chemicals by the electroreforming of biomass-derived alcohols, RSC Adv. 7 (2017) 13971−13978. Available from: https://doi.org/10.1039/c7ra00044h.
[41] M. Mougenot, A. Caillard, M. Simoes, S. Baranton, C. Coutanceau, P. Brault, PdAu/C catalysts prepared by plasma sputtering for the electro-oxidation of glycerol, Appl. Catal. B: Environ. 107 (2011) 372−379. Available from: https://doi.org/10.1016/j.apcatb.2011.07.039.
[42] A. Serov, U. Martinez, P. Atanassov, Novel Pd-In catalysts for alcohols electrooxidation in alkaline media, Electrochem. Commun. 34 (2013) 185−188. Available from: https://doi.org/10.1016/j.elecom.2013.06.003.
[43] A. Zalineeva, A. Serov, M. Padilla, U. Martinez, K. Artyushkova, S. Baranton, et al., Glycerol electrooxidation on self-supported $PdSn_x$ nanoparticules, Appl. Catal. B: Environ. 176 (2015) 429−435. Available from: https://doi.org/10.1016/j.apcatb.2015.04.037.
[44] A. Zalineeva, A. Serov, M. Padilla, U. Martinez, K. Artyushkova, S. Baranton, et al., Nano-structured Pd-Sn catalysts for alcohol electro-oxidation in alkaline medium, Electrochem. Commun. 57 (2015) 48−51. Available from: https://doi.org/10.1016/j.elecom.2015.05.006.
[45] A. Zalineeva, M. Padilla, U. Martinez, A. Serov, K. Artyushkova, S. Baranton, et al., Self-supported Pd-Bi catalysts for the electrooxidation of glycerol in alkaline media, J. Am. Chem. Soc. 136 (2014) 3937−3945. Available from: https://doi.org/10.1021/ja412429f.

[46] C. Coutanceau, A. Zalineeva, S. Baranton, M. Simoes, Modification of palladium surfaces by bismuth adatoms or clusters: effect on electrochemical activity and selectivity towards polyol electrooxidation, Int. J. Hydrogen Energy 39 (2014) 15877−15886. Available from: https://doi.org/10.1016/j.ijhydene.2014.03.076.
[47] A. Zalineeva, S. Baranton, C. Coutanceau, How do Bi-modified palladium nanoparticles work towards glycerol electrooxidation? An in situ FTIR study, Electrochim. Acta 176 (2015) 705−717. Available from: https://doi.org/10.1016/j.electacta.2015.07.073.
[48] Y. Holade, C. Morais, K. Servat, T.W. Napporn, K.B. Kokoh, Toward the electrochemical valorization of glycerol: Fourier transform infrared spectroscopic and chromatographic studies, ACS Catal. 3 (2013) 2403−2411. Available from: https://doi.org/10.1021/cs400559d.
[49] Y.X. Chen, A. Lavacchi, H.A. Miller, M. Bevilacqua, J. Filippi, M. Innocenti, et al., Nanotechnology makes biomass electrolysis more energy efficient than water electrolysis, Nat. Commun. 5 (2014) 4036. Available from: https://doi.org/10.1038/ncomms5036.
[50] M. Carmo, D.L. Fritz, J. Mergel, D. Stolten, A comprehensive review on PEM water electrolysis, Int. J. Hydrogen Energy 38 (2013) 4901−4934. Available from: https://doi.org/10.1016/j.ijhydene.2013.01.151.
[51] I.M. O'Hara, Z. Zhang, W.O. Doherty, C.M. Fellows, Lignocellulosics as a renewable feedstock for chemical industry: chemical hydrolysis and pretreatment processes, in: R. Sanghi, V. Singh (Eds.), Green Chemistry for Environmental Remediation, John Wiley & Sons Inc, Hoboken, 2011, pp. 505−560. Available from: http://dx.doi.org/10.1002/9781118287705.
[52] M. Benoit, A. Rodrigues, K. De Oliveira Vigier, E. Fourré, J. Barrault, J.M. Tatibouët, et al., Combination of ball-milling and non-thermal atmospheric plasma as physical treatments for the saccharification of microcrystalline cellulose, Green Chem. 14 (2012) 2212−2215. Available from: https://doi.org/10.1039/C2GC35710K.
[53] M. Benoit, A. Rodrigues, Q. Zhang, E. Fourré, K. De Oliveira Vigier, J.M. Tatibouët, et al., Depolymerization of cellulose assisted by a nonthermal atmospheric plasma, Angew. Chem. Int. Ed 50 (2011) 8964−8967. Available from: https://doi.org/10.1002/anie.201104123.
[54] Q. Zhang, M. Benoit, K. De Oliveira Vigier, J. Barrault, F. Jérôme, Pretreatment of microcrystalline cellulose by ultrasounds: effect of particle size in the heterogeneously-catalyzed hydrolysis of cellulose to glucose, Green Chem. 15 (2013) 963−969. Available from: https://doi.org/10.1039/C3GC36643J.
[55] Y.B. Vassilyev, O.A. Khazova, N.N. Nikolaeva, Kinetics and mechanism of glucose electrooxidation on different electrode-catalysts. Part I. Adsorption and oxidation on platinum, J. Electroanal. Chem. 196 (1985) 105−125. Available from: https://doi.org/10.1016/0022-0728(85)85084-1.
[56] F. Largeaud, K.B. Kokoh, B. Beden, C. Lamy, On the electrochemical reactivity of anomers: electrocatalytic oxidation of α and β-D-glucose on platinum electrodes in acid and basic media, J. Electroanal. Chem. 397 (1995) 261−269. Available from: https://doi.org/10.1016/0022-0728(95)04139-8.
[57] A. Brouzgou, P. Tsiakaras, Electrocatalysts for glucose electrooxidation reaction: a review, Top. Catal. 58 (2015) 1311−1327. Available from: https://doi.org/10.1007/s1124.
[58] K.B. Kokoh, J.M. Léger, B. Beden, C. Lamy, "On line" chromatographic analysis of the products resulting from the electrocatalytic oxidation of D-glucose on Pt, Au and adatoms modified Pt electrodes. Part I. Acid and neutral media, Electrochim. Acta 37 (1992) 1333−1342. Available from: https://doi.org/10.1016/0013-4686(92)87004-J.
[59] Spectral Database for Organic Compounds (SDBS), National Institute of Advanced Industrial Science and Technology (AIST-Japan). <http://sdbs.db.aist.go.jp/sdbs/cgi-bin/cre_index.cgi> (accessed 29.05.19).

[60] B. Beden, F. Largeaud, K.B. Kokoh, C. Lamy, Fourier transform infrared reflectance spectroscopic investigation of the electrocatalytic oxidation of d-glucose: identification of reactive intermediates and reaction products, Electrochim. Acta 41 (1996) 701−709. Available from: https://doi.org/10.1016/0013-4686(95)00359-2.
[61] Q. Wang, X. Cui, J. Chen, X. Zheng, C. Liu, T. Xue, et al., Well-dispersed palladium nanoparticles on graphene oxide as a non-enzymatic glucose sensor, RSC Adv. 2 (2012) 6245−6249. Available from: https://doi.org/10.1039/C2RA20425H.
[62] T.P. Tougas, M.J. Debenedetto, J.M. Demott, Postchromatographic electrochemical detection of carbohydrates at a silver-oxide electrode, Electroanalysis 5 (1993) 669−675. Available from: https://doi.org/10.1002/elan.1140050808.
[63] A. Stitz, W. Buchberger, Studies on electrochemical reactions at metal-oxide electrodes for combination with high performance liquid chromatography, Electroanalysis 6 (1994) 251−258. Available from: https://doi.org/10.1002/elan.1140060312.
[64] A. Cardoso de Sá, L. Lataro Paim, N. Ramos Stradiotto, Sugars electrooxidation at glassy carbon electrode decorate with multi-walled carbon nanotubes with nickel oxy-hydroxide, Int. J. Electrochem. Sci. 9 (2014) 7746−7762.
[65] M. Pasta, F. La Mantia, Y. Cui, Mechanism of glucose electrochemical oxidation on gold surface, Electrochim. Acta 55 (2010) 5561−5568. Available from: https://doi.org/10.1016/j.electacta.2010.04.069.
[66] Y. Holade, K. Servat, T.W. Napporn, C. Morais, J.-M. Berjeaud, K.B. Kokoh, Highly selective oxidation of carbohydrates in an efficient electrochemical energy converter: cogenerating organic electrosynthesis, ChemSusChem 9 (2016) 252−263.
[67] Y.X. Jiang, S.G. Sun, N. Ding, Novel phenomenon of enhancement of IR absorption of CO adsorbed on nanoparticles of Pd confined in supercages of Y-zeolite, Chem. Phys. Lett. 344 (2001) 463−470. Available from: https://doi.org/10.1016/S0009-2614(01)00812-0.
[68] M. Tominaga, M. Nagashima, K. Nishiyama, I. Taniguchi, Surface poisoning during electrocatalytic monosaccharide oxidation reactions at gold electrodes in alkaline medium, Electrochem. Commun. 9 (2007) 1892−1898. Available from: https://doi.org/10.1016/j.elecom.2007.04.024.
[69] S. Hermans, A. Deffernez, M. Devillers, Au−Pd/C catalysts for glyoxal and glucose selective oxidations, Appl. Catal. A: Gen. 395 (2011) 19−27. Available from: https://doi.org/10.1016/j.apcata.2011.01.019.
[70] L. Yan, A. Brouzgou, Y. Meng, M. Xiao, P. Tsiakaras, S. Song, Efficient and poison-tolerant Pd_xAu_y/C binary electrocatalysts for glucose electrooxidation in alkaline medium, Appl. Catal. B: Environ. 150−151 (2014) 268−274. Available from: https://doi.org/10.1016/j.apcatb.2013.12.026.
[71] T. Rafaïdeen, S. Baranton, C. Coutanceau, Highly efficient and selective electrooxidation of glucose and xylose in alkaline medium at carbon supported alloyed PdAu nanocatalysts, Appl. Catal. B: Environ. 243 (2019) 641−656.

CHAPTER SIX

Conclusions

Besides water, many other hydrogen-containing compounds, in particular organic compounds derived from biomass resources, can generate hydrogen through their electrochemical dissociation.

The electrochemical decomposition of water or of an organic compound gives hydrogen with much higher purity than that obtained by thermal processes such as steam reforming, autothermal reforming, or preferential oxidation. Moreover, there is no need to further clean-up the evolved gas, because no gaseous species, such as CO and CO_2, other than hydrogen are produced at the cathodic side of a proton exchange membrane electrolysis cell (PEMEC) thanks to the separation properties of the Proton Exchange Membrane. Water electrolysis is nearly a mature process, but it needs a relatively high amount of energy, c.5–6 kWh $(Nm^3\ H_2)^{-1}$, resulting from the high cell voltages used, that is, $U_{cell} \approx$ 1.5–2 V, due to high anodic overpotentials necessary for the oxygen evolution reaction (OER). This leads to high production cost not yet competitive with thermal processes, for example, steam methane reforming (SMR).

In this chapter, the feasibility of the production of clean hydrogen by the electrolysis of several organic compounds (formic acid, methanol, ethanol, glycerol, and glucose) in a PEMEC has been demonstrated with the following important features:

1. the quantity of hydrogen produced depends only on the electrolysis time Δt and the current intensity I, that is, on the quantity of electricity involved, $\Delta q = I\ \Delta t$, as shown by Eq. (2.13a) in Chapter 2, Section 2.2, as a linear function according to the Faraday's law, irrespective of the nature of the organic compound, of its concentration, of the electrode potential, of the electrocatalyst nature and structure and of the working temperature.
2. conversely the energy consumption W_e, proportional to the cell voltage $U_{cell}(I)$, depends greatly on the nature of the organic compound, its electrochemical reactivity, and eventually on its concentration in the case of electrode surface area non-saturated by adsorbed species, on the electrode potential, on the nature, composition, and structure

Production of Clean Hydrogen by Electrochemical Reforming of Oxygenated Organic Compounds.
DOI: https://doi.org/10.1016/B978-0-12-821500-5.00006-0

of the electrocatalyst and on the working temperature (electrochemical and thermal activations). The challenge is to find a convenient reactive organic compound and a suitable active electrocatalyst in order to reduce the overpotential, η_a, of the oxidation reaction, and thus to decrease the operational cell voltage, $U_{cell}(I)$, of the electrolysis process, that is, to decrease the quantity of electrical energy needed for producing hydrogen through its electrochemical reforming. The following important points have been underlined:

a. the favorable thermodynamic characteristics of the electrochemical decomposition of most organic compounds considered, leading to very low cell voltages under standard conditions, for example, $U^0_{MeOH} = 0.016$ V for methanol, compared to that of water ($U^0_{H2O} = 1.23$ V)—see Table 1 in Section 2.1.

b. depending on the organic compound considered, the nature, structure, and composition of the electrocatalyst are of prime importance to reduce the overpotential of the oxidation reaction and thus to decrease the cell voltage, for example, the following electrocatalysts give the best results:

  i. $Pd_{0.8}Au_{0.2}/C$ for the electrocatalytic reforming of formic acid;

  ii. $Pt_{0.5}Ru_{0.5}/C$ for methanol electrochemical reforming;

  iii. $Pt_{0.86}Sn_{0.1}Ru_{0.04}/C$ for ethanol electrochemical reforming;

  iv. $Pt_{0.5}Ru_{0.5}/C$ for glycerol electrooxidation in acidic media, and $Pt_{0.9}Bi_{0.1}/C$ or $Pd_{0.45}Pt_{0.45}Bi_{0.1}/C$ in alkaline media; and

  v. $Pt_{0.9}Sn_{0.1}/C$ in acidic media or $Pd_{0.3}Au_{0.7}/C$ in alkaline media for glucose electrochemical reforming.

c. the electrocatalytic activity of these electrocatalysts strongly depend on their structure, for example, the particle size and preferential orientation of nanoparticles dispersed on a carbon powder such as Vulcan XC72 so that their synthesis method must be thoroughly controlled, for example, the Bönnemann method for the preparation of catalysts for methanol and ethanol electrooxidation, or the water-in-oil micro-emulsion method in the case of formic acid, glycerol, and glucose oxidation.

d. the development of more active and stable electrocatalysts for the electrochemical reforming of organic compounds is required to decrease further the anodic overpotential of the oxidation reaction. The best approach to meet this challenge is to acquire a detailed knowledge of the reaction mechanism, particularly the identification of the rate determining step (rds), the rate of which must be increased through the development of new electrocatalysts. The investigation of reaction mechanisms needs to use complementary physical methods such as electrochemical methods (voltammetry, chronoamperometry, etc.) coupled with physicochemical (*in situ* Infrared Reflectance Spectroscopy, Differential Electrochemical Mass Spectroscopy) and analytical methods (gas or liquid chromatography *online*).

e. the increase of the working temperature has always had a positive effect by increasing the reaction rate, that is, the current intensity at a given cell voltage, by thermal activation. As an example, at $U_{cell} = 600$ mV for the electrooxidation of 2 M methanol the current density increases from 10 mA cm^{-2} at 25°C to 43.5 mA cm^{-2} at 85°C.

f. in most cases, the electrochemical reforming of these organic compounds inside a PEM electrolyzer occurs at cell voltages (U_{cell} < 1 V, i.e., W_e < 2 kWh $(Nm^3)^{-1}$ H_2) lower than those of water electrolysis (1.5 V < U_{cell} < 2 V, i.e., W_e ≈ 4 to 6 kWh $(Nm^3)^{-1}$ H_2), allowing to reduce greatly, at least by a factor of 2–4, the electrical energy consumed.

3. for hydrogen production from electrochemical reforming of alcohol, the energy efficiency at 25°C, assuming a complete oxidation to CO_2, decreases as follows: methanol (30%) > ethanol (18%), > glycerol (estimated to 7%). These energy efficiencies are at least two times smaller than those encountered in water electrolysis (60%–80%), but this is not a key point since W_e is much smaller, and the biomass resources are available in a much greater amount.

But as the complexity of the reaction process increases with the number of carbon atoms, the hydrogen yield, as defined by the ratio of the number of hydrogen moles produced per one mole of the reactive compound encountering total oxidation to CO_2, decreases in the order: formic acid and methanol (1/1 = 3/3 = 100%) > ethanol (2/6 = 33%) > glycerol (1/7 = 14%)—see Table 2.1 in Chapter 2, Section 2.1.

The last point to underline is the fact that it seems actually difficult to produce hydrogen by the electrochemical reforming of biomass compounds with more than two carbon atoms. But the electrolysis of such compounds, for example, glycerol or glucose investigated here, may lead to high value-added chemicals with a lower cost of production.

A final conclusion is that any organic compound will lead to similar results, since the rate of hydrogen evolution only depends on ñ, the average number of Faradays per mole of hydrogen involved in the overall process ($\tilde{n} = 2$ for any organic compound). Thus the most interesting organic compounds will be those experiencing a lower cell voltage for their electrochemical decomposition and a high yield of hydrogen produced, which requires a right choice of the best efficient electrocatalyst leading to the lowest overpotential at the highest current intensity, that is, a high activity, and to the most complete oxidation reaction to CO_2 if possible, that is, a high selectivity under the working conditions of the electrolysis process.

Index

Note: Page numbers followed by "*f*" and "*t*" refer to figures and tables, respectively

FSC
www.fsc.org
MIX
Papier | Fördert gute Waldnutzung
FSC® C083411

Kolibri 360 GmbH
Ferdinand-Jühlke-Straße 7
99095 Erfurt, Deutschland
produktsicherheit@kolibri360.de